T0212221

Customizable Computing

Synthesis Lectures on Computer Architecture

Editor
Margaret Martonosi, *Princeton University*

Synthesis Lectures on Computer Architecture publishes 50- to 100-page publications on topics pertaining to the science and art of designing, analyzing, selecting and interconnecting hardware components to create computers that meet functional, performance and cost goals. The scope will largely follow the purview of premier computer architecture conferences, such as ISCA, HPCA, MICRO, and ASPLOS.

Customizable Computing
Yu-Ting Chen, Jason Cong, Michael Gill, Glenn Reinman, and Bingjun Xiao
2015

Single-Instruction Multiple-Data Execution
Christopher J. Hughes
2015

Die-stacking Architecture
Yuan Xie and Jishen Zhao
2015

Power-Efficient Computer Architectures: Recent Advances
Magnus Själander, Margaret Martonosi, and Stefanos Kaxiras
2014

FPGA-Accelerated Simulation of Computer Systems
Hari Angepat, Derek Chiou, Eric S. Chung, and James C. Hoe
2014

A Primer on Hardware Prefetching
Babak Falsafi and Thomas F. Wenisch
2014

On-Chip Photonic Interconnects: A Computer Architect's Perspective
Christopher J. Nitta, Matthew K. Farrens, and Venkatesh Akella
2013

Optimization and Mathematical Modeling in Computer Architecture
Tony Nowatzki, Michael Ferris, Karthikeyan Sankaralingam, Cristian Estan, Nilay Vaish, and David Wood
2013

Security Basics for Computer Architects
Ruby B. Lee
2013

The Datacenter as a Computer: An Introduction to the Design of Warehouse-Scale Machines, Second edition
Luiz André Barroso, Jimmy Clidaras, and Urs Hölzle
2013

Shared-Memory Synchronization
Michael L. Scott
2013

Resilient Architecture Design for Voltage Variation
Vijay Janapa Reddi and Meeta Sharma Gupta
2013

Multithreading Architecture
Mario Nemirovsky and Dean M. Tullsen
2013

Performance Analysis and Tuning for General Purpose Graphics Processing Units (GPGPU)
Hyesoon Kim, Richard Vuduc, Sara Baghsorkhi, Jee Choi, and Wen-mei Hwu
2012

Automatic Parallelization: An Overview of Fundamental Compiler Techniques
Samuel P. Midkiff
2012

Phase Change Memory: From Devices to Systems
Moinuddin K. Qureshi, Sudhanva Gurumurthi, and Bipin Rajendran
2011

Multi-Core Cache Hierarchies
Rajeev Balasubramonian, Norman P. Jouppi, and Naveen Muralimanohar
2011

A Primer on Memory Consistency and Cache Coherence
Daniel J. Sorin, Mark D. Hill, and David A. Wood
2011

Chip Multiprocessor Architecture: Techniques to Improve Throughput and Latency
Kunle Olukotun, Lance Hammond, and James Laudon
2007

Transactional Memory
James R. Larus and Ravi Rajwar
2006

Quantum Computing for Computer Architects
Tzvetan S. Metodi and Frederic T. Chong
2006

Customizable Computing

Yu-Ting Chen, Jason Cong, Michael Gill, Glenn Reinman, and Bingjun Xiao

ISBN: 978-3-031-00620-3 paperback
ISBN: 978-3-031-01748-3 ebook

DOI 10.1007/978-3-031-01748-3

A Publication in the Springer series
SYNTHESIS LECTURES ON ADVANCES IN AUTOMOTIVE TECHNOLOGY

Lecture #33
Series Editor: Margaret Martonosi, *Princeton University*
Series ISSN
Print 1935-3235   Electronic 1935-3243

Customizable Computing

Yu-Ting Chen, Jason Cong, Michael Gill, Glenn Reinman, and Bingjun Xiao
University of California, Los Angeles

SYNTHESIS LECTURES ON COMPUTER ARCHITECTURE #33

ABSTRACT

Since the end of Dennard scaling in the early 2000s, improving the energy efficiency of computation has been the main concern of the research community and industry. The large energy efficiency gap between general-purpose processors and application-specific integrated circuits (ASICs) motivates the exploration of customizable architectures, where one can adapt the architecture to the workload. In this Synthesis lecture, we present an overview and introduction of the recent developments on energy-efficient customizable architectures, including customizable cores and accelerators, on-chip memory customization, and interconnect optimization. In addition to a discussion of the general techniques and classification of different approaches used in each area, we also highlight and illustrate some of the most successful design examples in each category and discuss their impact on performance and energy efficiency. We hope that this work captures the state-of-the-art research and development on customizable architectures and serves as a useful reference basis for further research, design, and implementation for large-scale deployment in future computing systems.

KEYWORDS

accelerator architectures, memory architecture, multiprocessor interconnection, parallel architectures, reconfigurable architectures, memory, green computing

Contents

Acknowledgments

This research is supported by the NSF Expeditions in Computing Award CCF-0926127, by C-FAR (one of six centers of STARnet, an SRC program sponsored by MARCO and DARPA), and by the NSF Graduate Research Fellowship Grant #DGE-0707424.

Yu-Ting Chen, Jason Cong, Michael Gill, Glenn Reinman, and Bingjun Xiao
June 2015

CHAPTER 1

Introduction

Since the introduction of the microprocessor in 1971, the improvement of processor performance in its first thirty years was largely driven by the Dennard scaling of transistors [45]. This scaling calls for for reduction of transistor dimensions by 30% every generation (roughly every two years) while keeping electric fields constant everywhere in the transistor to maintain reliability (which implies that the supply voltage needs to be reduced by 30% as well in each generation). Such scaling doubles the transistor density each generation, reduces the transistor delay by 30%, and at the same time improves the power by 50% and energy by 65% [7]. The increased transistor count also leads to more architecture design innovations, such as better memory hierarchy designs and more sophisticated instruction scheduling and pipelining supports. These factors combined led to over 1,000 times performance improvement of Intel processors in 20 years (from the $1.5\mu m$ generation down to the 65 nm generation), as shown in [7].

Unfortunately, Dennard scaling came to an end in the early 2000s. Although the transistor dimension reduction by 30% per generation continues to follow Moore's law, the supply voltage scaling had to almost come to a halt due to the rapid increase of leakage power. In this case, transistor density can continue to increase, but so can the power density. As a result, in order to continue meeting the ever-increasing computing needs, yet maintaining a constant power budget, in the past ten years the computing industry stopped simple processor frequency scaling and entered the era of parallelization, with tens to hundreds of computing cores integrated in a single processor, and hundreds to thousands of computing servers connected in a warehouse-scale data center. However, such highly parallel, general-purpose computing systems now face serious challenges in terms of performance, power, heat dissipation, space, and cost, as pointed out by a number of researchers. The term "utilization wall" was introduced in [128], where it shows that if the chip fills up with 64-bit adders (with input and output registers) designed in a 45 nm TSMC process technology running at the maximum operating frequency (5.2Ghz in their experiment), only 6.5% of $300mm^2$ of the silicon can be active at the same time. This utilization ratio drops further to less than 3.5% in the 32nm fabrication technology, roughly by a factor of two in each technology generation following their leakage-limited scaling model [128].

A similar but more detailed and realistic study on dark silicon projection was carried out in [51]. It uses a set of 20 representative Intel and AMD cores to build up empirical models which capture the relationship between area vs. performance and the relationship between power vs. performance. These models, together with the device-scaling models, are used for projection of the core area, performance, and power in various technology generations. This also considers real parallel application workloads as represented by the PARSEC benchmark suite [9]. It further considers different multicore models, including the symmetric multicores, asymmetric multicores (consisting of both large and small cores), dynamic multicores (either large or small cores depending on if the power or area constraint is imposed), and composable multicores (where small cores can be fused into large cores). Their study concludes that at 22 nm, 21% of a fixed-size chip must be powered off, and at 8 nm, this dark silicon ratio grows to more than 50% [51]. This study also points to the end of simple core scaling.

Given the limitation of core scaling, the computing industry and research community are actively looking for new disruptive solutions beyond parallelization that can bring further significant energy efficiency improvement. Recent studies suggest that the next opportunity for significant power-performance efficiency improvement comes from customized computing, where one may adapt the processor architecture to optimize for intended applications or application domains [7, 38].

The performance gap between a totally customized solution using an application-specific integrated circuit (ASIC) and a general-purpose processor can be very large, as documented in several studies. An early case study of the 128-bit key AES encryption algorithm was presented in [116]. An ASIC implementation of this algorithm in a $0.18\mu m$ CMOS technology achieves a 3.86Gbits/second processing rate at 350mW power consumption, while the same algorithm coded in assembly languages yields a 31Mbits/second processing rate with 240mW power running on a StrongARM processor, and a 648Mbits/second processing rate with 41.4W power running on a Pentium III processor. This results in a performance/energy efficiency (measured in Gbits/second/W) gap of a factor of 85X and 800X, respectively, when compared with the ASIC implementation. In an extreme case, when the same algorithm is coded in the Java language and executed on an embedded SPARC processor, it yields 450bits/second with 120mW power, resulting in a performance/energy efficiency gap as large as a factor of 3 million (!) when compared to the ASIC solution.

Recent work studied a much more complex application for such gap analysis [67]. It uses a 720p high-definition H.264 encoder as the application driver, and a four-core CMP system using the Tensilica extensible RISC cores [119] as the baseline processor configuration. Compared to an optimized ASIC implementation, the baseline CMP is 250X slower and consumes 500X more energy. Adding 16-wide SIMD execution units to the baseline cores improves the performance by 10X and energy efficiency by 7X. Addition of custom-fused instructions is also considered, and it improves the performance and energy efficiency by an additional 1.4X. Despite these enhancements, the resulting enhanced CMP is still 50X less energy efficient than ASIC.

The large energy efficiency gap between the ASIC and general-purpose processors is the main motivation for architecture customization, which is the focus of this lecture. In particular, one way to significantly improve the energy efficiency is to introduce many special-purpose on-chip accelerators implemented in ASIC and share them among multiple processor cores, so that as much computation as possible is carried out on accelerators instead of using general-purpose cores. This leads to accelerator-rich architectures, which have received a growing interest in recent years [26, 28, 89]. Such architectures will be discussed in detail in Chapter 4.

There are two major concerns about using accelerators. One relates to their low utilization and the other relates to their narrow workload coverage. However, given the utilization wall [128] and the dark silicon problem [51] discussed earlier, low accelerator utilization is no longer a serious problem, as only a fraction of computing resources on-chip can be activated at one time in future technology generation, given the tight power and thermal budgets. So, it is perfectly fine to populate the chip with many accelerators, knowing that many of them will be inactive at any given time. But once an accelerator is used, it can deliver one to two orders of magnitude improvement in energy efficiency over the general-purpose cores.

The problem of narrow workload coverage can be addressed by introducing reconfigurability and using composable accelerators. Examples include the use of fine-grain field-programmable gate arrays (FPGAs), coarse-grain reconfigurable arrays [61, 62, 91, 94, 118], or dynamically composable accelerator building blocks [26, 27]. These approaches will be discussed in more detail in Section 4.4.

Given the significant energy efficiency advantage of accelerators and the promising progress in widening accelerator workload coverage, we increasingly believe that the future of processor architecture should be rich in accelerators, as opposed to having many general-purpose cores. To some extent, such accelerator-rich architectures are more like a human brain, which has many specialized neural microcircuits (accelerators), each dedicated to a different function (such as navigation, speech, vision, etc.). The high degree of customization in the human brain leads to a great deal of efficiency; the brain can perform various highly sophisticated cognitive functions while consuming only about 20W, an inspiring and challenging performance for computer architects to match.

Not only can the compute engine be customized, but so can the memory system and on-chip interconnects. For example, instead of only using a general-purpose cache, one may use program-managed or accelerator-managed buffers (or scratchpad memories). Customization is needed to flexibly partition these two types of on-chip memories. Memory customization will be discussed in Chapter 5. Also, instead of using a general-purpose mesh-based network-on-chip (NoC) for packet switching, one may prefer a customized circuit-switching topology between accelerators and the memory system. Customization of on-chip interconnects will be discussed in Chapter 6.

The remainder of this lecture is organized as follows. Chapter 2 gives a broad overview of the trajectory of customization in computing. Customization of compute cores, such as custom instructions, will be covered in Chapter 3. Loosely coupled compute engines will be discussed in Chapter 4. Chapter 5 will discuss customizations to the memory system, and Chapter 6 discusses custom network interconnect designs. Finally, Chapter 7 concludes the lecture with discussions of industrial trends and future research topics.

CHAPTER 2

Road Map

Customized computing involves the specialization of hardware for a particular domain, and often includes a software component to fully leverage this specialization in hardware. In this section, we will lay the foundation for customized computing, enumerating the design trade-offs and defining vocabulary.

2.1 CUSTOMIZABLE SYSTEM-ON-CHIP DESIGN

In order to provide efficient support of customized computing, the general-purpose CMP (chip multiprocessor) widely used today needs to be replaced or transformed into a Customizable System-on-a-Chip (CSoC), also called customizable heterogeneous platform (CHP) in some other publications [39], which can be customized for a particular domain through the specialization of four major components on such a computing platform, including: (1) processor cores, (2) accelerators and co-processors, (3) on-chip memory components, and (4) the network-on-chip (NoC) that connects various components. We will explore each of these in detail individually, as well as in concert with the other CSoC components.

2.1.1 COMPUTE RESOURCES

Compute components like processor cores handle the actual processing demands of the CSoC. There are a wide array of design choices in the compute components of the CSoC. But when looking at customized compute units, there are three major factors to consider, all of which are largely independent of one another:

- **Programmability**

- **Specialization**

- **Reconfigurability**

Programmability

A *fixed function* compute unit can do one operation on incoming data, and nothing else. For example, a compute unit that is designed to perform an FFT operation on any incoming data is fixed function. This inflexibility limits how much a compute unit may be leveraged, but it streamlines the design of the unit such that it may be highly optimized for that particular task. The amount of bits used within the datapath of the unit and the types of mathematical operators included for example can be precisely tuned to the particular operation the compute unit will perform.

Contrasting this, a programmable compute unit executes sequences of instructions to define the tasks they are to perform. The instructions understood by the programmable compute unit constitute the instruction set architecture (ISA). The ISA is the interface for use of the programmable compute unit. Software that makes use of the programmable compute unit will consist of these instructions, and these instructions are typically chosen to maximize the expressive nature of the ISA to describe the nature of computation desired in the programmable unit. The hardware of the programmable unit will handle these instructions in a generally more flexible datapath than that of the fixed function compute unit. The fetching, decoding, and sequencing of instructions leads to performance and power overhead that is not required in a fixed function design. But the programmable compute unit is capable of executing different sequences of instructions to handle a wider array of functions than a fixed function pipeline.

There exists a broad spectrum of design choices between these two alternatives. Programmable units may have a large number of instructions or a small number of instructions for example. A pure fixed function compute unit can be thought of as a programmable compute unit that only has a single implicit instruction (i.e., perform an FFT). The more instructions supported by the compute unit, the more compact the software needs to be to express desired functionality. The fewer instructions supported by the compute unit, the simpler the hardware required to implement these instructions and the more potential for an optimized and streamlined implementation. Thus the programmability of the compute unit refers to the degree to which it may be controlled via a sequence of instructions, from fixed function compute units that require no instructions at all to complex, expressive programmable designs with a large number of instructions.

Specialization

Customized computing targets a smaller set of applications and algorithms within a domain to improve performance and reduce power requirements. The degree to which components are customized to a particular domain is the specialization of those components. There are a large number of different specializations that a hardware designer may utilize, from the datapath width of the compute unit, to the number of type of functional units, to the amount of cache, and more.

This is distinct from a general purpose design, which attempts to cover all applications rather than providing a customized architecture for a particular domain. General purpose designs may use a set of benchmarks from a target performance suite, but the idea is not to optimize specifically for those benchmarks. Rather, that performance suite may simply be used to gauge performance.

There is again a broad spectrum of design choices between specialized and general purpose designs. One may consider general purpose designs to be those specialized for the domain of all applications. In some cases, general purpose designs are more cost effective since the design time may be amortized over more possible uses—an ALU that can be designed once and then used in a variety of compute units may amortize the cost of the design of the ALU, for example.

Reconfigurability

Once a design has been implemented, it can be beneficial to allow further adaptation to continue to customize the hardware to react to (1) changes in data usage patterns, (2) algorithmic changes or advancements, and (3) domain expansion or unintended use. For example, a compute unit may have been optimized to perform a particular algorithm for FFT, but a new approach may be faster. Hardware that can flexibly adapt even after tape out is reconfigurable hardware. The degree to which hardware may be reconfigured depends on the granularity of reconfiguration. While finer granularity reconfiguration can allow greater flexibility, the overhead of reconfiguration can mean that a reconfigurable design will perform worse and/or be less energy efficient than a static (i.e., non-reconfigurable) alternative. One example of a fine-grain reconfigurable platform is an FPGA, which can be used to implement a wide array of different compute units, from fixed function to programmable units, with all levels of specialization. But an FPGA implementation of a particular compute unit is less efficient than an ASIC implementation of the same compute unit. However, the ASIC implementation is static, and cannot adapt after design tape out. We will examine more coarse-grain alternatives for reconfigurable compute units in Section 4.4.

Examples

- Accelerators—early GPUs, mpeg/media decoders, crypto accelerators
- Programmable Cores—modern GPUs, general purpose cores, ASIPs
- Future designs may feature accelerators in primary computational role
- Some programmable cores and or programmable fabric are still included for generality/longevity

Section 3 covers the customization of processor cores and Section 4 covers coprocessors and accelerators. We split compute components into two sections to better manage the diversity of the design space for these components.

2.1.2   ON-CHIP MEMORY HIERARCHY

Chips are fundamentally pin-limited, which impacts the amount of bandwidth that can be supplied to the compute units described in the previous section. This is further exacerbated by limitations in DRAM scaling. On-chip memory is one technique to mitigate this. On-chip memory can be used in a variety of ways, from providing data buffering for streaming data from off-chip to providing a place to store data that will be reused multiple times for computation. Once again, different applications will have different on-chip memory requirements. As with compute units, there are a wide array of design choices for hardware architects to consider in the design of the memory hierarchy.

Transparency to Software

A cache is relatively small, but fast memory which leverages the principle of locality to reduce the latency to access memory. Data which will be reused in the near future is kept in the cache to avoid accesses to longer latency memory. There are two primary approaches to managing a cache (i.e., orchestrating what data comes into the cache and what data leaves the cache): purely hardware approaches and software managed caches. In this book, we will use the term *scratchpad* to refer to software-based caches, where the application writer or the compiler will be responsible for explicitly bringing data into and out of the cache through special instructions. Hardware caches where actual control circuits orchestrate data movement without software intervention will just be referred to as caches in this book. Scratchpads have tremendous potential for application-specific customization since cache management can be tuned to a particular application, but they also come with coding overhead as the programmer or compiler writer must explicitly map out this orchestration. Conventional caches are more flexible, as they can handle a wider array of applications without requiring explicit management, and may be preferable in cases where the access pattern is unpredictable and therefore requires dynamic adaptation.

Sharing

On-chip memory may be kept private to a particular compute unit or may be shared among multiple compute units. Private on-chip memory means that the application will not need to contend for space with another application, and will get the full benefit of the on-chip memory. Shared on-chip memory can amortize the cost of on-chip memory over several compute units, providing a potentially larger pool of space that can be leveraged by these compute units than if the space was partitioned among the units as private memory. For example, four compute units can each have 1MB of on-chip memory dedicated to them. Each compute unit will always have 1MB regardless of the demand from other compute units. However, if four compute units share 4MB of on-chip memory, and if the different compute units use different amounts of memory, one compute unit may, for example, use more than 1MB of space at a particular time since there is a large pool of memory available. Sharing works particularly well when compute units use different amounts of memory at different times. Sharing is also extremely effective when compute units make use of the same memory locations. For example, if compute units are all working on an image in parallel, storing the image in a single memory shared among the units allows the compute units to more effectively cooperate on the shared data.

2.1.3   NETWORK-ON-CHIP

On-chip memory stores the data needed by the compute units, but an important part of the over-all CSoC is the communication infrastructure that allows this stored data to be distributed to the compute units, that allows the data to be delivered to/from the on-chip memory from/to the memory interfaces that communicate off-chip, and that allows compute units to synchronize and communicate with one another. In many applications there is a considerable amount of data that must be communicated to the compute units used to accelerate application performance. And with multiple compute units often employed to maximize data level parallelism, there are often multiple data streams being communicated around the CSoC. These requirements transcend the conventional bus-based design of older multicore designs, with designers instead choosing network-on-chip (NoC) designs. NoC designs enable the communication of more data between more CSoC components.

Components interfacing with an NoC typically bundle transmitted data into packets, which contain at least address information as to the desired communication destination and the payload itself, which is some portion of the data to be transmitted to a particular destination. NoCs transmit messages via packets to enable flexible and reliable data transport—packets may be buffered at intermediate nodes within the network or reordered in some situations. Packet-based communication also avoids long latency arbitration that is associated with communication in a single hop over an entire chip. Each hop through a packet-based NoC performs local arbitration instead.

The creation of an NoC involves a rich set of design decisions that may be highly customized for a set of applications in a particular domain. Most design decisions impact the latency or bandwidth of the NoC. In simple terms, the latency of the NoC is how long it takes a given piece of data to pass through the NoC. The bandwidth of the NoC is how much data can be communicated in the NoC at a particular time. Lower latency may be more important for synchronizing communication, like locks or barriers that impact multiple computational threads in an application. Higher bandwidth is more important for applications with streaming computation (i.e., low data locality) for example.

One example of a design decision is the topology of an NoC. This refers to the pattern of links that connect particular components of the NoC. A simple topology is a *ring*, where each component in the NoC is connected to two neighboring components, forming a chain of components. More complex communication patterns may be realized by more highly connected topologies that allow more simultaneous communication or a shorter communication distance.

Another example is the bandwidth of an individual link in the topology—wire that is traversed in one cycle of the network's clock. Larger links can improve bandwidth but require more buffering space at intermediate network nodes, which can increase power cost.

An NoC is typically designed with a particular level of utilization in mind, where decisions like topology or link bandwidth are chosen based on an expected level of service. For example, NoCs may be designed for worst case behavior, where the bandwidth of individual links is sized for peak traffic requirements, and every path in the network is capable of sustaining that peak bandwidth requirement. This is a flexible design in that the worst case behavior can manifest on any particular communication path in the NoC, and there will be sufficient bandwidth to handle it. But it can mean overprovisioning the NoC if worst case behavior is infrequent or sparsely exhibited in the NoC. In other words, the larger bandwidth components can mean wasted power (if only static power) or area in most cases. NoCs may also be designed for average case behavior, where the bandwidth is sized according to the average traffic requirement, but in such cases performance can suffer when worst case behavior is exhibited.

Topological Customization

Customized designs can specialize different parts of the NoC for different communication patterns seen in applications within a domain. For example, an architecture may specialize the NoC such that there is a high bandwidth connection between a memory interface and a particular compute unit that performs workload balancing and sorting for particular tasks, and then there is a lower bandwidth connection between that compute unit for workload balancing and the remainder of the compute units that perform the actual computation (i.e., work). More sophisticated designs can adapt bandwidth to the dynamic requirements of the application in execution. Customized designs may also adapt the topology of the NoC to the specific requirements of the application in execution. Section 6.2 will explore such flexible designs, along with some of the complexity in implementing NoC designs that are specialized for particular communication patterns.

Routing Customization

Another approach to specialization is to change the routing of packets in the NoC. Packets may be scheduled in different ways to avoid congestion in the NoC, for example. Another example would be *circuit switching*, where a particular route through the NoC is reserved for a particular communication, allowing packets in that communication to be expedited through the NoC without intermediate arbitration. This is useful in bursty communication where the cost of arbitration may be amortized over the communication of many packets.

Physical Design Customization

Some designs leverage different types of wires (i.e., different physical trade-offs) to provide a heterogeneous NoC with specialized communication paths. And there are also a number of exciting alternative interconnects that are emerging for use in NoC design. These alternative interconnects typically improve interconnect bandwidth and reduce communication latency, but may require some overhead (such as upconversion to an analog signal to make use of the alternative interconnect). These interconnects have some physical design and architectural challenges, but also provide some interesting options for customized computing, as we will discuss in Section 6.4.

2.2   SOFTWARE LAYER

Customization is often a holistic process that involves both hardware customization and software orchestration. Application writers (i.e., domain experts) may have intimate knowledge of their applications which may not be expressed easily or at all in traditional programming languages. Such information could include knowledge of data value ranges or error tolerance, for example. Software layers should provide a sufficiently expressive language for programmers to communicate their knowledge of the applications in a particular domain to further customize the use of specialized hardware.

There are a number of approaches to programming domain-specific hardware. A common approach is to create multiple layers of abstraction between the application programmer and the domain-specific hardware. The application programmer writes code in a relatively high level language that is expressive enough to capture domain-specific information. The high level language uses library routines implemented in the lower levels of abstraction as much as possible to cover the majority of computational tasks. The library routines may be implemented through further levels of abstraction, but ultimately lead to a set of primitives that directly leverage domain-specific hardware. As an example, library routines to do FFTs may leverage hardware accelerators specifically designed for FFT. This provides some portability of the higher level application programmer code, while still providing domain-specific specialization at the lower abstraction levels that directly leverages customized hardware. This also hides the complexity of customized hardware from the application writer.

Another question is how much of the process of software mapping may be automated. Compilers may be able to perform much of the mapping of high level code to customized hardware through intelligent algorithms that can transform code and leverage application-specific information from the application programmer. Automation is a powerful tool for discovering opportunities for acceleration in code that may not be covered by existing library routines.

CHAPTER 3

Customization of Cores

3.1 INTRODUCTION

Because processing cores contribute greatly to energy consumption in modern processors, the conventional processing core is a good place to start looking for customizations to computation engines. Processing cores are pervasive, and their architecture and compilation flow are mature. Modifications made to processing cores then have the advantage that existing hardware modules and infrastructure invested in building efficient and high-performance processors can be leveraged, without having to necessarily abandon existing software stacks as may be required when designing hardware from the ground up. Additionally, programmers can use their existing knowledge of programming conventional processing cores as a foundation toward learning new techniques that build upon conventional cores, instead of having to adopt new programming paradigms, or near languages.

In addition to benefiting from mature software stacks, any modifications made to a conventional processing core can also take advantage of many of the architectural components that have made cores so effective. Examples of these architectural components are caches, mechanisms for out-of-order scheduling and speculative execution, and software scheduling mechanisms. By integrating modifications directly into a processing core, new features can be designed to blend into these components. For example, adding a new instruction to the existing execution pipeline automatically enables this instruction to benefit from aggressive instruction scheduling already present in a conventional core.

However, introducing new compute capability, such as new arithmetic units, into existing processing cores means being burdened by many of the design restrictions that these cores already exert on arithmetic unit design. For example, out-of-order processing benefits considerably from short latency instructions, as long latency instructions can cause pipeline stalls. Conventional cores are also fundamentally bound, both in terms of performance and efficiency, by the infrastructure necessary to execute instructions. As a result, conventional cores cannot be as efficient at performing a particular task as a hardware structure that is more specialized to that purpose [26]. Figure 3.1 illustrates this point, showing that the energy cost of executing an instruction is much greater than the energy that is required to perform the arithmetic computation (e.g., energy devoted to integer and floating point arithmetic). The rest of the energy is spent to implement the infrastructure internal to the processing core that is used to perform tasks such as scheduling instructions, fetch and decode, extracting instruction level parallelism, etc. Figure 3.1 shows only the comparison of structures internal to the processing core itself, and excludes exter-

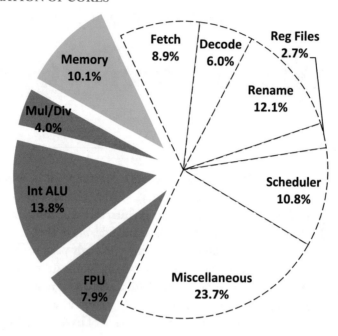

Figure 3.1: Energy consumed by subcomponents of a conventional compute core as a proportion of the total energy consumed by the core. Subcomponents that are not computationally necessary (i.e., they are part of the architectural cost of extracting parallelism, fetching and decoding instructions, scheduling, dependency checking, etc.) are shown as slices without fill. Results are for a Nehalem era 4-core Intel Xeon CPU. Memory includes L1 cache energy only. Taken from [26].

nal components such as memory systems and networks. These are burdens that are ever present in conventional processing cores, and they represent the architectural cost of generality and programmability. This can be contrasted against the energy proportions shown in Figure 3.2, which show the energy saving when the compute engine is customized for a particular application, instead of a general-purpose design. The difference in energy cost devoted to computation is primarily the result of relaxing the design requirements of functional units, so that functional units operate only at precisions that are necessary and are designed to emphasize energy efficiency per computation, and potentially exhibit deeper pipelines and longer latencies than would be tolerable when couched inside a conventional core.

This chapter will cover the following topics related to customization of processing cores:

- **Dynamic Core Scaling and Defeaturing:** A post-silicon method of selectively deactivating underutilized components with the goal of conserving energy.

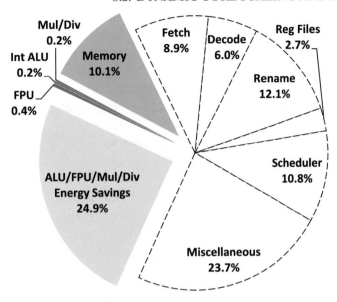

Figure 3.2: Energy cost of subcomponents in a conventional compute core as a proportion of the total energy consumed by the core. This shows the energy savings attainable if computation is performed in an energy-optimal ASIC. Results are for a Nehalem era 4-core Intel Xeon CPU. Memory includes L1 cache energy only. Taken from [26].

- **Core Fusion:** Architectures that enable one "big" core to act as if it were really many "small cores," and vice versa, to dynamically adapt to different amounts of thread-level or instruction-level parallelism.

- **Customized Instruction Set Extensions:** Augmenting processor cores with new workload-specific instructions.

3.2   DYNAMIC CORE SCALING AND DEFEATURING

When a general-purpose processor is designed, it is done with the consideration of a wide range of potential workloads. For any particular workload, many resources may not be fully utilized. As a result, these resources continue to consume power, but do not contribute meaningfully to program performance. In order to improve energy efficiency architectural features can be added that allow for these components to be selectively turned off. While this obviously does not allow for the chip area spent on deactivated components to be repurposed, it does allow for a meaningful energy efficiency improvement.

Manufacturers of modern CPUs enable this type of selective defeaturing, though typically not for this purpose. This is done with the introduction of machine-specific registers that indicate the activation of particular components. The purpose of these registers, from a manufacturer's perspective, is to improve processor yield by allowing faulty components in an otherwise stable processor to be disabled. For this reason, the machine-specific registers governing device activation are rarely documented.

There has been extensive academic work in utilizing defeaturing to create dynamically heterogeneous systems. These works center around identifying when a program is entering a code region that systemically underutilizes some set of features that exist in a conventional core. For example, if it is possible to statically discover that a code region contains long sequences of dependencies between instructions, then it is clear that a processor with a wide issue and fetch width will not be able to find enough independent instructions to make effective use of those wide resources [4, 8, 19, 125]. In that case, powering off the components that enable wide fetch and issue, along with the architectural support for large instruction windows, can save energy without impacting performance. This academic work is contingent upon being able to discern run-time behavior or code, either using run-time monitoring [4, 8] or static analysis [125].

An example of dynamic resource scaling from academia is CoolFetch [125]. CoolFetch relies on compiler support to statically estimate the execution rate of a code region, and then uses this information to dynamically grow and contract structures within a processor's fetch and issue units. By constraining these structures in code regions with few opportunities to exploit instruction-level parallelism or out-of-order scheduling, CoolFetch also observed a carry-over effect of reducing power consumption of other processor structures that normally operate in parallel, and also reduces energy spent on squashed instructions by reducing reducing the number of instructions stalling at retirement. In all, CoolFetch reported on average 8% energy savings, with a relatively trivial architectural modification and a negligible performance penalty.

3.3   CORE FUSION

The efficiency of a processing core, both in terms of energy consumption and compute per area, tends to be reduced as the potential performance of the core increases. The primary cause of this is a shifting of focus from investing resources in compute engines in the case of small cores, to aggressive scheduling mechanisms in the case of big cores. In modern out-of-order processors, this scheduling mechanism constitutes the overwhelming majority of core area investment, and the overwhelming majority of energy consumption.

An obvious conclusion then is that large sophisticated cores are not worth including in a system, since a sea of weak cores provides greater potential for system-wide throughput than a small group of powerful cores. The problem with this conclusion, however, is that parallelizing software is difficult: parallel code is prone to errors like race conditions, and many algorithms are limited by sequential components that are more difficult to parallelize. Some code cannot reasonably be parallelized at all. In fact, the majority of software is not parallelized at all, and thus cannot make use of a large volume of cores. In these situations a single powerful core is preferable, since it offers high single-thread throughput at the cost of restricting the capability to exploit thread-level parallelism. Because of this observation, it becomes clear that the best design depends on the number of threads that are exposed in software. Large numbers of threads can be run on a large number of cores, enabling higher system-wide throughput, while a few threads may be better run on a few powerful cores, since multiple cores cannot be utilized.

This observation gave rise to a number of academic works that explore heterogeneous systems which feature a small group of very powerful cores on die with a large group of very efficient cores [64, 71, 84]. In addition to numerous academic perspectives on heterogeneous systems, industry has begun to adopt this trend, such as the ARM big.LITTLE [64]. While these designs are interesting, they still allocate compute resources statically, and thus cannot react to variation in the degree of parallelism present in software. To address this rigidity, core fusion [74], and other related work [31, 108, 115, 123], propose mechanisms for implementing powerful cores out of collaborative collections of weak cores. This allows a system to grow and shrink so that there are as many "cores" in the system as there are threads, and each of these cores is scaled to maximize performance with the current amount of parallelism in the system.

Core fusion [74] accomplishes this scaling by splitting a core into two halves: a narrow-issue conventional core with the fetch engine stripped off, and an additional component that acts as a modular fetch/decode/commit component. This added component will either perform fetches for each core individually from individual program sequences, or a wide fetch to feed all processors. Similar to how a line buffer reads in multiple instructions in a single effort, this wide fetch engine will read an entire block of instructions and issue them across different cores. Decode and resource renaming is also performed collectively, with registers being stored as physically resident in various cores. A crossbar is added to move register values from one core to another when necessary. At the end of the pipeline, a reordering step is introduced to guarantee correct commit and exception handling. A diagram of this architecture is shown in Figure 3.3. Two additional instructions are added to this architecture that allow the operating system to merge and split core collections, thus adjusting the number of virtual cores available for scheduling.

As shown in Figure 3.4, Core Fusion cores perform only slightly worse than a natural processor with the same issue width, achieving performance within 20% of a monolithic processor at equivalent effective issue width. The main reason for this is that the infrastructure that enables fusion comes with a performance cost. The strength of this system, however, is its adaptability, not necessarily its performance when compared to a processor designed for a particular software configuration. Furthermore, energy required to power structures necessary for wide out-of-order scheduling do not need to be active when cores are not fused. As a result, core fusion surrenders a portion of area used to implement the out-of-order scheduler and about 20% performance when fused to emulate a larger core. This enables run-time customization of core width and number, in a way that is binary compatible, and thus is completely transparent. For systems that do not have a priori knowledge of the type of workload that will be running on a processor, or expect the software to transition between sequential and parallel portions, the ability to adjust and accommodate varying workloads is a great benefit.

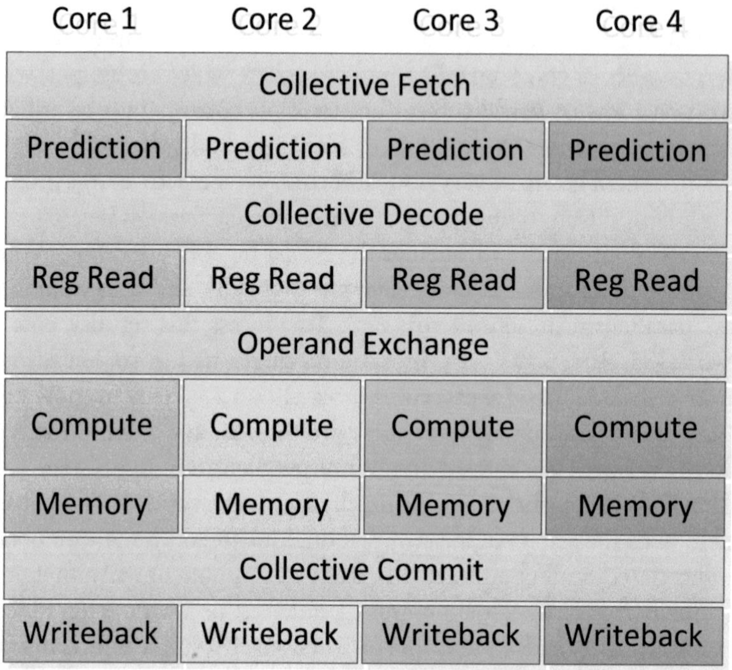

Core 1	Core 2	Core 3	Core 4
Collective Fetch			
Prediction	Prediction	Prediction	Prediction
Collective Decode			
Reg Read	Reg Read	Reg Read	Reg Read
Operand Exchange			
Compute	Compute	Compute	Compute
Memory	Memory	Memory	Memory
Collective Commit			
Writeback	Writeback	Writeback	Writeback

Figure 3.3: A 4-core core fusion processor bundle with components added to support merging of cores. Adapted from [74].

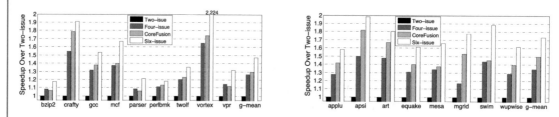

Figure 3.4: Comparison of performance between various processors of issue widths and 6-issue merged core fusion. Taken from [74].

3.4   CUSTOMIZED INSTRUCTION SET EXTENSIONS

In a conventional, general-purpose processor design, each time an instruction is executed, it must pass through a number of stages of a processor pipeline. Each of these stages incurs a cost, which is dependent on the type of processor. Figure 3.1 showed the energy consumed in various stages of the processor pipeline. In terms of the core computational requirement of an application, the energy spent in the execute stage is energy spent doing productive compute work, and everything else (i.e., instruction fetch, renaming, instruction window allocation, wakeup and select logic) is overhead required to support and accelerate general-purpose instruction processing for a particular architecture. The reason for execution constituting such a small portion of energy consumed is that for most instructions, each performs a small amount of work.

Extending the instruction set of an otherwise conventional compute core to increase the amount of work done per instruction is one way of improving both performance and energy efficiency for particular tasks. This is accomplished by merging the tasks that would have otherwise been performed by multiple instructions, into a single instruction. This is valuable because this single large instruction still only requires a single pass through the fetch, decode, and commit phases, and thus requires a reduced amount of bookkeeping to be maintained to perform the same task. In addition to reducing the overhead associated with processing an instruction, ISA extensions enable access to custom compute engines to implement these composite operations more efficiently than could be implemented otherwise.

The strategy of instruction set customization ranges from very simple (e.g., [6, 95, 111]) to complex (e.g., [63, 66]). Simplistic but effective instruction set extensions are now common in commodity processors in the form of specialized vector instructions: SSE and AVX instructions. Section 3.4.1 discusses vector instructions, which allow for simple operations, mostly floating point operations, to be packed into a single instruction and operate over a large volume of data, potentially simultaneously. While these vector instructions are restricted to use in regular, compute-dense code, they lend a large enough performance advantage that processor manufacturers are continuing to push toward more feature-rich vector extensions [55].

In addition to vector instructions, there has also been work proposed by both industry [95] and academia [63] that ties multiple operations together into a single compute engine that operates over a single element of data. These custom compute engines are discussed in Section 3.4.2, and differ from vector instructions in that they describe a group of operations over a small set of data, rather than the reverse. Thus, they can be tied more tightly into the critical path of a conventional core [136].

While fixed-functionality custom compute engines offer an opportunity for chip designers to design optimal engines, they are inflexible, and thus prone toward low utilization. This has motivated work in programmable instruction set extension [63]. Configurable custom instructions are instead implemented as a programmable datapath coupled with a portion of memory that holds a configuration for the datapath. These programmable extensions are described in Section 3.4.3.

3.4.1   VECTOR INSTRUCTIONS

A conceptually simple, but very effective avenue for instruction set customization is the introduction of vector instructions, also known as Single Instruction–Multiple Data [SIMD] instructions. These perform a single, primitive operation on a volume of data. Vector instructions are introduced typically with two architectural components: an array of new registers that store vectors of arguments for the purpose of acting as arguments for vector instructions, and an array of compute engines that perform the parallel computation of vector instructions.

The concept of vector computation and vector instructions is an old topic. In spite of this, vector instructions continue to see active research because they are highly effective at communicating a large amount of computation using very few instructions. Vector instructions have also seen acceptance in virtually all commodity processors. In the case of x86 & x86-64 processors, which dominate consumer processors, vector instructions come in the form of SSE instructions [111] and AVX instructions [55]. These instructions operate over small vectors of between 4 and 16 elements, mostly for the purpose of performing floating-point arithmetics. In academia there has been a substantial amount of work on vector processing [5, 52, 68], frequently with a focus on vector sizes of a much larger size compared to what has been seen in commodity processors, such as observed in the Tarantula architecture [52] which features 512 element vectors.

In order to further improve performance, many vector architectures feature a lane-based design, which is to say that there are multiple small compute engines and register elements that participate in parallel to perform a computation. Each lane consists of a set of compute engines, and a portion of a register file corresponding to a subset of vector elements. Figure 3.5 illustrates this design. Each engine computes a subset of the entire computation, and multiple engines run simultaneously. Issuing a vector instruction results in an instruction issued to all lanes at once, even though renaming and dependency tracking can be performed per vector, rather than per element. An increase in the number of lanes reduces the work performed by a single engine, thus increasing performance, while fewer lanes result in reduced area consumption at the cost of performance. An ideal number of lanes typically is dependent on available memory bandwidth, as it is still necessary to load and store vector data, and improving compute capability ultimately results in an increase in memory stalls.

3.4.2   CUSTOM COMPUTE ENGINES

An alternative form of extending an instruction set is to introduce instructions that merge multiple primitive operations on a single set of data into a single instruction. An example of this that appears frequently in commodity processors is a single instruction that performs a floating point multiply and add in a single instruction, referred to as a fused multiply-add [95], which is a common pattern in linear algebra applications.

The architectural task of introducing custom instructions during the design of a processor is relatively straightforward. The majority of academic work has instead focused on identification of what custom compute engines should be included as instruction extensions (e.g., [2, 24, 25, 136]), or compilation strategies to improve mapping of program code to custom instructions (e.g., [25]). While these engines are not necessarily reconfigurable, they are customizations for a set of applications.

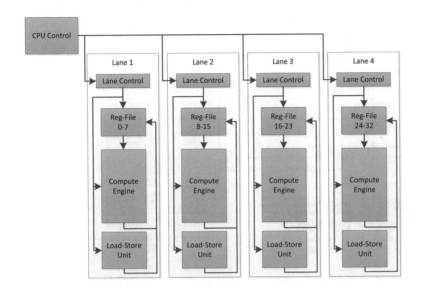

Figure 3.5: A vector architecture with individual compute lanes.

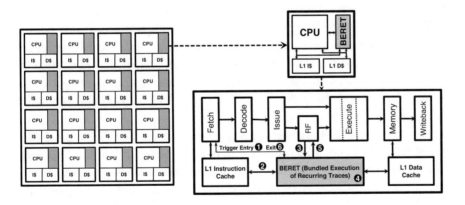

Figure 3.6: How a BERET module fits into a processor pipeline. Taken from [66].

3.4.3 RECONFIGURABLE INSTRUCTION SETS

Reconfigurable instruction set architectures allow for a program to author its own instructions [63, 66]. An example of a configurable custom instruction architecture is BERET [66]. The architecture for BERET, which stands for Bundled Execution of Recurring Traces, is shown in Figure 3.6 and Figure 3.7, and consists of programmable compute engines called subgraph ex-

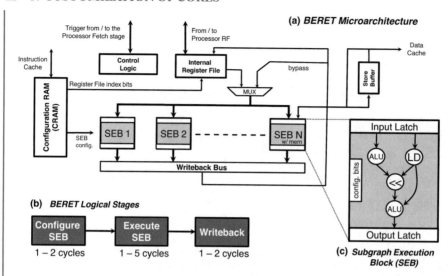

Figure 3.7: Internal architecture and use of BERET modules. (a) Shows internal BERET architecture along with communication mechanisms with SEBs, (b) shows approximate overhead associated with different stages of a BERET invocation, (c) shows a configuration stored in a SEB. Taken from [66].

ecution blocks (SEBs) and configuration memory to hold trace configurations. Each trace configuration holds a small set of pre-decoded instructions that are tightly scheduled. A user program uses two instructions to interact with configurable instructions: 1) a command to program the configuration memory, and 2) a command to invoke a stored trace. To simplify integration into the rest of the core pipeline, a BERET engine is restricted to working on a subset of registers that would normally be available, thus limiting the number of inputs and outputs, and is restricted from issuing memory accesses directly.

Figure 3.8 shows the process of transforming a program to use a BERET engine, which is a process performed statically at compile time. A program is first separated into "hot code," which is frequently executed code where the majority of time is spent, and "cold code," which is the large volume of code that is executed infrequently (Figure 3.8.a). The hot code region is then broken up into communicating sub-regions, on the granularity of a basic block (Figure 3.8.b). These sub-regions are then converted into BERET SEB configurations, each of which describes a small program segment with no control dependencies (Figure 3.8.c) which can be loaded and invoked at runtime. The hot code region is then expressed in terms of dependent BERET calls, and possibly some flow control being executed within the unmodified portions of the processing core. At run time, these configurations are loaded into the SEB units, and the hot code is executed by invoking the BERET engine (Figure 3.8.d). Tasks like dependency need only be performed when transitioning from one BERET call to another, and can be performed by the mechanisms that normally perform instruction dependency tracking within the unmodified processing core.

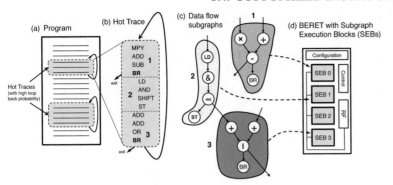

Figure 3.8: Program transformation process to map software to a BERET module. A hot region is first broken down into chunks, with each chunk stored as a SEB configuration. The core dependency tracking is used to schedule inter-SEB activities. Taken from [66].

This simple architecture allows the BERET engine to perform the majority of the work associated with executing a compute kernel, and leaves management of control structures and memory operations for the general-purpose portion of the core.

While a configurable architecture like BERET is not going to execute a particular functionality as efficiently or as fast as a dedicated compute engine, the configurable nature of the engine allows for high utilization while executing hot code, without being restricted to particular types of operations. It also accomplishes the primary goal of customized instruction sets, which allows resource utilization to be further tilted toward the execution stage of the pipeline and away from other stages, and increases the amount of work performed per instruction.

3.4.4   COMPILER SUPPORT FOR CUSTOM INSTRUCTIONS

Compiler technology that assists the use of custom instructions falls broadly into two categories: 1) pattern matching in a data-flow graph to map custom compute engines [2, 24, 25, 136], and 2) loop analysis to automate the application of vector instructions [49, 50, 82].

Pattern matching at its core involves solving a subgraph isomorphism problem [44], which is proven to be NP-Complete. In spite of this, there has been much work in making this process tractable for small patterns [25], which has done much to make pattern matching reasonable in practice. Custom instruction design tends to focus on small patterns, typically on the order of 2 to 10 instructions, both to increase utilization and to make the task of pattern matching possible. Pattern matching is also attractive because it is a very local optimization in many cases, without requiring any complex supporting analysis. A compiler author can simply apply a pattern-matching routine on any arbitrary program region.

Automating vectorization is a more complicated topic and has been researched extensively for decades [49, 50, 82]. Vectorization primarily focuses around loop bound analysis and data layout manipulation, both of which require more complicated analyses than simple pattern matching, along with detailed alias analysis, which is itself a separate and complex topic. Because automatic vectorization requires much more knowledge about runtime state, such as resolved alias analysis and known factors for runtime variables, vectorization is a problem that is comparably difficult to solve, and often requires a great deal of inter-process analysis to be most effective.

CHAPTER 4

Loosely Coupled Compute Engines

4.1 INTRODUCTION

In Chapter 3 we looked at customizable compute resources that were tied to a conventional core. For example, a processor might be augmented with hardware to support a customized instruction. In this chapter we look at customized hardware that is not tied to a particular core: these acceleration engines instead are used to offload computation from a conventional core and compute it in a more efficient manner. These components are shared among multiple cores and thus multiple thread contexts to amortize the cost of this specialized hardware. These disparate compute resources may then be asynchronous with respect to one another, forming a set of heterogeneous resources that individually execute parts of a program for which they have been specialized. Programs structured for this environment are no longer a single logical stream of discrete commands, but instead are a fluid collaboration of different tasks that may execute on specialized hardware for that particular task.

Making a jump to this model of computing comes with some advantages and disadvantages. Stand-alone engines can circumvent components that a conventional core would benefit from, if those components would hamper the computation being performed. For example, compute engines used for streaming style computations can circumvent caches, long latencies are less impactful if a compute engine isn't tied to a processor pipeline, and speculative prefetchers are not necessary if a compute engine can seemlessly tolerate long latencies associated with memory accesses. The primary disadvantage of splitting a computation between a conventional processor and an accelerator is that software must be designed to run across heterogeneous devices, and to tolerate asynchronous communication between these devices. From a software perspective, this is comparable to requiring that software be designed in a multi-threaded way. From a hardware perspective though, this enables opportunities for introducing specialized functionality without having to tie it to an individual processing core, and instead have a core simply borrow a compute engine for the duration it is needed, instead of owning it outright.

Ultimately, the potential for increased performance and energy efficiency that comes with deeper specialization and relaxation of design constraints, and the potential to leverage compute resources distributed among a complete system, make loose coupling between conventional cores and specialized accelerators very attractive for a variety of compute-intensive tasks. Couple this with the practical fact that accelerators can typically be designed and tested in isolation, without having to redesign and retest a complete compute core, and it is clear why loosely coupled compute engines are an attractive route for improving compute performance and efficiency.

While it is an interesting topic in its own right, this chapter will not cover PCI devices or other off-chip devices, or compute engines that are conventionally introduced into a system in the form of an off-chip device. The main purpose of this is to focus the content of this chapter more on the topic of emerging research, and away from topics that have already seen pervasive adoption. This primarily means the exclusion of graphics processing units (GPUs), whether or not they are integrated into the CPU, which are already discussed in many other books, and have long been a common fixture in both high-performance computing platforms and consumer machines. Also excluded are the plethora of ASIC microcontrollers present in commodity machines which do not interact with software, such as Ethernet controllers, USB controllers, conventional disk controllers, etc.

This chapter will cover the following topics related to non-core compute engines:

- **Loosely Coupled Accelerators** Coarse-grain compute engines that act autonomously from processing cores.

- **Field Programmable Fabrics** Ultra-fine-grain reconfigurability that sacrifices some efficiency and performance for generality.

- **Coarse-Grain Reconfigurable Arrays** Composable accelerators with near-ASIC performance and FPGA-like configurability.

4.2   LOOSELY COUPLED ACCELERATORS

Loosely coupled accelerators (LCAs) are coarse-grain compute engines that exist as disjoint devices attached to an interconnect instead of being affixed to a specific core. While these devices can be either on-chip or off-chip, this section will focus on on-chip LCAs.

Accelerators have been examined both in academia [28, 112] and in industry [58, 98] as an effective way to greatly improve efficiency and performance for mature codes. By introducing an accelerator as an LCA that is not physically adjoined to a specific processing core, it can potentially be shared by all cores in a system. This allowed for a potential increase in compute engine utilization, and thus more efficient use of area dedicated to the LCA. A drawback of this decoupling, however, is that there is latency associated with invoking an LCA. This added latency means that the LCA must perform a larger amount of work to justify the cost of invoking it. This typically takes the form of operating over a larger volume of data rather than performing a computation that is more intensive. As a result of this, LCAs also typically feature a simple control mechanism that allows them to iterate over a volume of data, such as a DMA engine, that can be configured to target large buffers for input and output. By absorbing a small control structure into the compute engine, an LCA can operate for many thousands of cycles autonomously, thus making the comparably large start-up cost affordable.

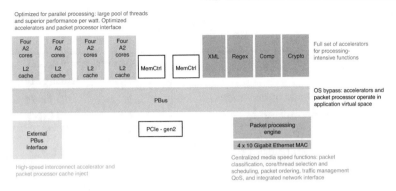

Figure 4.1: A diagram of the WSP processor. Taken from [58].

A significant shortcoming of LCAs in general is that they are implemented as ASIC, and are thus of fixed functionality. This restricts LCAs to only being reasonably used in instances in which an algorithm is (1) mature enough that it is unlikely to change in the near future and (2) important enough that there is justification for creating a specialized processor. While there are a set of workloads that meet these criteria, such as encryption and decryption for web servers, workloads that are worth the up-front cost associated with designing a special-purpose processor are infrequent. While tapeout costs are dropping for processors, and the tools for processor authorship are improving, it is still difficult to economically justify the inclusion of LCAs in commodity processors in most cases.

4.2.1 WIRE-SPEED PROCESSOR

The Wire-Speed Processor (WSP) from IBM [58], an influential platform shown in Figure 4.1, is an example of a processor featuring LCAs. This is a particularly interesting example of an LCA featuring architecture for two reasons: 1) this work states clear performance requirements, which are beyond what can be offered from conventional processors, and 2) because WSP accelerators were considered a central feature, rather than a device that facilitates computation behind the scenes. A WSP is expected to feature a large library of LCAs that are selected based on their value to facilitate computation within a very specific domain, and thus a WSP is not expected to be general-purpose, but instead is highly customized at the time of manufacturing. The selection of accelerators is not configurable at run time, but the selection of which LCAs to include is based on which programs are expected to run on the WSP.

As the name would indicate, a WSP is designed with a heavy emphasis on high throughput, and so it provides accelerators that are designed to offer performance far beyond what equivalent software would be able to offer. The WSP also features cores to act as control mechanisms between accelerator invocations. Table 4.1 shows throughput targets for LCAs included in the original WSP work.

Table 4.1: Performance of accelerators included in original WSP work [58]. Note that achieved bandwidth is aggregate across multiple engines. Taken from [58]

Accelerator unit	Functionality	No. engines	Typical throughput	Peak throughput
HEA	Network node mode	4	40 Gb/s	40 Gb/s
	Endpoint mode	4	40 Gb/s	40 Gb/s
Compression	Gzip (input bandwidth)	1	8 Gb/s	16 Gb/s
	Gunzip (output bandwidth)	1	8 Gb/s	16 Gb/s
Encryption	AES	3	41 Gb/s	60 Gb/s
	TDES	8	19 Gb/s	60 Gb/s
	ARC4	1	5.1 Gb/s	60 Gb/s
	Kasumi	1	5.9 Gb/s	60 Gb/s
	SHA	6	23-37 Gb/s	60 Gb/s
	MD5	6	31 Gb/s	60 Gb/s
	AES/SHA	3	19-31 Gb/s	60 Gb/s
	RSA & ECC (1024-bit)	3	45 Gb/s	60 Gb/s
XML	Customer workload	4	10 Gb/s	30 Gb/s
	Benchmark	4	20 Gb/s	30 Gb/s
Regex	Average pattern sets	8	20-40 Gb/s	70 Gb/s

Architectures like that found in the WSP are highly customized for a particular workload or set of workloads. While this allows them to achieve performance and efficiency far beyond that which would normally be found in a conventional processor for these workloads, they are indistinguishable from a resource-constrained conventional processor for workloads other than those for which they are customized.

4.2.2   COMPARING HARDWARE AND SOFTWARE LCA MANAGEMENT

Sharing LCAs among multiple cores comes with an added arbitration cost. Arbitration can be either performed by a system driver or by adding a specialized hardware mechanism [28]. There are benefits and drawbacks to both of these strategies.

Exposing LCAs to software by a device driver is a conventional way of performing arbitration. A driver in this case would guarantee that access to any given LCA is only done by a single software thread context at a time. A driver can be an arbitrarily sophisticated piece of software, and thus can implement an arbitrarily sophisticated scheduling policy. In instances where scheduling is very important, this could potentially be very useful. It should be considered though that the introduction of a driver increases the cost associated with invoking an accelerator, since interacting with a device driver is potentially a very expensive process.

Introducing a dedicated hardware mechanism for interacting with LCAs potentially reduces the overhead associated with invoking an accelerator, but limits the potential sophistication of scheduling [28]. The overhead reduction is particularly important if the expectation is that the communication protocol between the controlling core and the LCA involves a large volume of communication. Introducing a hardware mechanism also allows the protection that a device driver would normally provide to instead be provided entirely in user-space code [28], which eliminates costly transitions to privileged execution modes.

4.2.3   UTILIZING LCAS

Programmatically finding uses for an LCA in general program code brings with it all the challenges discussed in Section 3.4.4, only with large patterns instead of small patterns. The problem however is identical, and can be approached with the same solution: solving subgraph isomorphism. But, because LCAs generally consist of relatively large numbers of operations, sometimes thousands of instructions, the subgraph isomorphism problem rapidly becomes intractable. In addition to this, because the large graphs implemented by LCAs can often be expressed in many different ways, automating discovery of LCA targetable regions very likely requires additionally proving program equivalence between enumerated subgraphs and the implemented LCA. In all, expecting a compiler to make effective use of a large LCA is currently unreasonable.

Instead of relying on a compiler, LCA-featuring systems typically rely on an API that is provided to software authors. This is the strategy taken by all works cited and discussed in this section. This API provides a clean interface to interacting with an LCA, and exposes all features that the suite of LCAs can offer. While the use of this API requires that a program be written to target a particular platform, it allows programmers access to additional compute resources that lend performance and energy benefits well beyond what could be achieved otherwise. This benefit is enticing enough for some programmers to adopt a new API in migrating to a particular platform with LCAs, but widespread adoption is hampered by the lack of effective compiler-based support for LCA deployment.

4.3   ACCELERATORS USING FIELD PROGRAMMABLE GATE ARRAYS

The previous section discussed coarse-grain LCAs implemented as ASIC as one extreme end of the spectrum of runtime programmability and flexibility. These LCAs provide area and energy efficient performance, but are inflexible. Accelerators implemented using field programmable gate array (FPGA) fabric are the opposite extreme, offering enormous flexibility at the cost of efficiency and performance.

FPGA fabric consists of a large matrix of extremely primitive compute elements, sitting on top of a matrix of circuit-switched interconnect components. The entire structure is configurable, and can be used to implement any arbitrary circuit, which in turn implements an arbitrary functionality. The primary components that constitute the compute elements of an FPGA fabric are lookup tables (LUTs), which accept as input a series of single-bit signals, and index into a result table for values to forward on as output signals. In addition to these compute elements, an FPGA fabric will also contain two primitive memory elements: 1) flip-flops that act as latch registers between pipeline stages, and 2) Block RAMs (BRAMs) which are small ASIC memory cells for addressable local storage. These components, suspended in a sea of circuit-switched networking material, allow for the communication of virtually anything that could be produced as an ASIC circuit.

The degree of flexibility that FPGAs offer, however, comes with a set of steep costs. First, configuration of these devices is relatively time-consuming and involves moving a considerable amount of memory, since a configuration describing every tiny component of the circuit needs to be loaded from memory. Second, designing the components to be loaded becomes a synthesis problem. This can take days of computation to translate the design into an appropriate configuration, and thus has a dramatic impact on compilation time. In spite of this, FPGAs have become a frequent target for a wide range of acceleration applications, both in the academic space and in industry where an ASIC system would be produced in insufficient volume to financially justify silicon tapeout.

Prior to the maturation of high level synthesis [35, 54] (HLS), programming an FPGA was enormously difficult and time-consuming when compared to the effort involved in writing conventional software. This high development cost has made FPGAs a choice of last resort for many of those who end up adopting them, in spite of their high performance potential and low energy consumption. HLS changed this by providing a programmatic way of describing an FPGA design by converting conventional program code to a hardware description language like VHDL, thus enabling rapid development and easing the process of adapting conventional software to make use of FPGA fabric. This, combined with advances in compiler technology to enable automated extraction of program regions suitable for migration to FPGA fabric, are making FPGAs usable as part of a conventional tool chain. This is shifting FPGAs from their conventional role as platforms on which to implement fully customized embedded designs to the role of platforms that introduce acceleartors alongside conventional processors in heterogeneous systems.

The first commercially available step in this direction was the Xilinx Zynq line, which featured a set of processors alongside an FPGA, all of which were cache coherent. This has allowed for the exploration of heterogeneous systems in new ways [23].

4.4 COARSE-GRAIN RECONFIGURABLE ARRAYS

Section 4.2 and Section 4.3 discussed the two extremes, LCAs and FPGAs: extreme performance and efficiency on one end, and extreme reconfiguration on the other. Coarse-grain reconfigurable arrays (CGRAs), or composable accelerators, attempt to bridge the gap by offering a set of coarse-grain components and a mechanism to network them together [26, 92, 104]. The granularity of these compute engines is large enough to capture much of the circuit-level efficiency and performance found in LCAs, while offering most of the useful programmability and reconfigurability of FPGAs. This is done by coarsening the granularity of components being networked together from individual LUTs and registers into whole functional units that may each perform several operations. Additionally, because there are relatively few components being networked together, there is a higher ratio of compute material to networking material than that found in an FPGA, and more opportunity to tolerate more dynamic mapping methods than could be tolerated in an FPGA. Conceptually, a CGRA platform can be thought of as an FPGA with small accelerators instead of LUTs, whether or not a given architecture is implemented to match this analogy.

Conceptually, a CGRA architecture can be thought of as a set of custom instructions, as described in Chapter 3.4, suspended in a network. Each of these small compute engines is only responsible for implementing a small amount of computation, similar in scope to that which would be performed by a handful of instructions if they were executed on a normal compute core, and then passing the results on to the next compute engine. The next compute engine would perform a few more operations, then pass its output to the next compute engine. Together, a communicating graph of these compute engines perform a complex operation, with each individual compute engine in this graph performing a small part.

This sort of design comes with several benefits. First, individual compute engines can be implemented very efficiently, instead of repeatedly implementing common compute engines out of ultra-fine-grain primitives, such as LUTs. This allows for much greater compute density than can be achieved in FPGA fabric, while achieving near-ASIC performance and energy efficiency. Second, because an accelerator is composed of relatively few components, the process of mapping these components to hardware resources is substantially less complicated. Instead of mapping tens or hundreds of thousands of components together, and producing networks between them, a CGRA configuration might involve the networking of a dozen components, or even fewer. This not only speeds compilation, but also potentially speeds configuration, and thus reduces the overhead associated with using a CGRA accelerator as compared to an FPGA accelerator. A third benefit is improved utilization as compared to a fully customized ASIC LCA, which comes about because individual compute engines can be used in a wide variety of situations. This flexibility allows for a larger portion of a program to be covered by CGRA accelerators, and thus improves the overall impact of introducing a CGRA system, even if the benefits of a CGRA platform are less than that of a pure ASIC implementation on any individual program kernel.

To act as a driving example for use in this section, consider the 16-point multiply-accumulate operation shown in Figure 4.2(a). For the sake of argument, suppose that an individual compute engine, unless otherwise specified, performs up to three arithmetic operations. Thus, as shown in Figure 4.2(b), the sample equation can be broken up into five pieces which must communicate with one another to perform the complete operation that is desired.

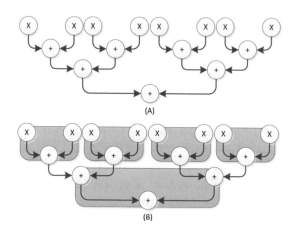

Figure 4.2: A sample equation to act as a driver for further discussion. (a) Shows the full computation to be performed, (b) shows the computation broken into work to be performed by individual compute engines.

4.4.1   STATIC MAPPING

A number of CGRA works [81, 104, 118] have focused on a static mapping of resources, where compute engines and the communication between these compute engines is calculated either off-line or as part of a compilation pass. This sort of design allows for a very simple network design, such as the circuit-switched networking infrastructure typically seen internal to FPGA fabric, to be the communication medium between compute engines. Because individual compute engines potentially operate at a high frequency, there isn't tolerance for irregular latencies in networks. For this reason, the pattern is mapped to the CGRA substrate as it is specified at compile time, where the timing of the communication between compute engines may be considered. Figure 4.3 shows an example of a statically mapped CGRA architecture [118].

A common feature of CGRA platforms is the virtualization of accelerator resources. This is done by specifying an accelerator as a communicating set of compute engines by their positions relative to one another. A resource management mechanism then looks for a match for the requested pattern [81]. This allocation strategy is clearly biased toward the use of homogeneous compute engines, which is commonly seen in proposed statically mapped CGRA designs [104]. There are also designs that advocate homogeneous clusters of heterogeneous compute engines [81].

When mapping the example program shown in Figure 4.2, the particular architecture used will influence what communication patterns are possible. Consider an example CGRA platform with circuit-switched links connecting neighbors in all directions (horizontally, vertically, and diagonally) in a 3x4 compute engine grid. Figure 4.4(a) shows a labeling of nodes in our sample

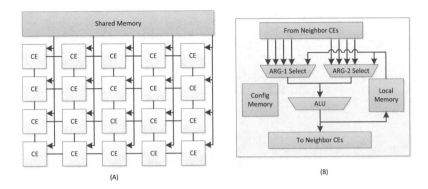

Figure 4.3: An example CGRA example, inspired by [118]. (a) Shows the interconnection of multiple compute engines, (b) shows the internal structure of a single compute engine.

compute graph, with Figure 4.4(b) showing a potential mapping of these nodes to compute engines such that our desired communication pattern can be achieved. At run time, we can select from any arrangement of compute engine such that the pattern shown in Figure 4.4(b) can fit, and so that the compute engines are not otherwise occupied with other tasks. One such allocation of our sample 3x4 CGRA is shown in Figure 4.4(c), though there are four possible configurations to choose from (shifting the pattern to the right, up, or to the upper right). At compile time the pattern shown in Figure 4.4(b) would be generated, while the allocation shown in Figure 4.4(c) would be performed at run time. This example also illustrates the importance of homogeneity in the CGRAs layout, as an unusable compute engine anywhere in the center of the sample CGRA would eliminate all possible mapping opportunities.

4.4.2 RUN-TIME MAPPING

Since CGRA platforms consist of few communicating components, with each component being responsible for a large portion of work, it is possible for a CGRA platform to use a packet-switched or dynamic network between compute engines instead of a static circuit switched network [26]. This complicates matters as the design must tolerate fluctuations in latency, but it allows for a higher degree of dynamism in system design and accelerator utilization. In these systems it is not necessary to know where compute engines are relative to one another, which trivializes accelerator virtualization, because any compute engine can communicate with any other compute engine. This design also greatly simplifies the task of scheduling accelerators when it is assumed that accelerator resources will be contested. Figure 4.5 illustrates the difference in perspective between the compilation and mapping components targeting a run-time mapped CGRA, and the actual architecture.

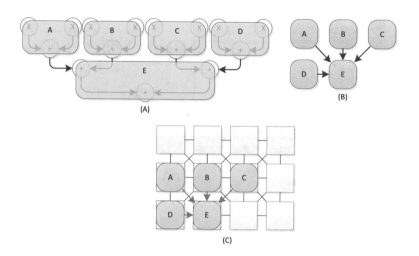

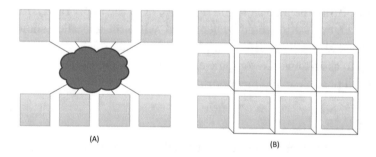

Figure 4.4: The procedure of mapping a sample program to a statically mapped sample CGRA architecture. (a) Shows the broken up program to be mapped with node labels, (b) shows a sample communication pattern that can be used to implement this computation, (c) shows a sample allocation of this mapping to a sample 3x4 CGRA with neighbor links.

Figure 4.5: A CGRA architecture designed for mapping at runtime with a packet switched network. (a) Shows how the compiler & mapping system views the CGRA (as fully connected), with (b) illustrating the actual topological layout.

Using a packet-switched network requires a greater demand for resources, as compared to using a static circuit-switched network, both in terms of area use and energy efficiency. However, because this network is only used to connect coarse-grain components together, the overhead observed by transitioning to a packet-switched system does not have the impact that would be observed if, for example, an FPGA featured a packet-switched network between components. A larger cost associated with this design is the infrastructure required to tolerate latency fluctuation. Each compute engine in a packet-switched CGRA system requires buffering to accommodate irregularities in communication latency, and handshaking is required between communicating pairs of devices, along with control structures that allow a compute engine to tolerate stalls across sequences of communicating compute engines.

While resource allocation can be performed more aggressively in a run-time mapped CGRA, the algorithm used for compute engine allocation has a high impact on the actual observed performance of the composed accelerator. A statically mapped accelerator can be mapped into whatever its optimal construction is, and force the resource management system to either grant this optimal decision or fail outright. In a system that tolerates a more flexible communication pattern, accelerators of varying quality can be constructed from the same compute engines, based on the network traffic created from the use of the composed accelerator. While this further increases possible resource utilization, it also eliminates a guarantee of the quality of a composed accelerator.

System design for a CGRA platform that supports arbitrary communication between resources is also simplified. In order to compile a program, the compiler need only have an understanding of a library of potential compute engines available for use, and a guaranteed minimum quantity of each kind of compute engine. The physical orientation of these compute engines relative to one another doesn't determine which compositions are viable, but instead determines the performance of composed accelerators. For this reason, a system designer has greater freedom with regard to designing a system consisting of a more varied library of compute engines, since introducing heterogeneity into a CGRA system no longer impacts compiler decisions or validity of mapping decisions. Knowledge of the orientation of compute engines in a run-time mapped CGRA only becomes important to the mechanism performing resource management, and then only for the purpose of estimating performance impact.

To illustrate the difference run-time mapping makes, consider again the sample program shown in Figure 4.2. In this case, because the view of the CGRA from the perspective of the mapping routine is effectively as a fully connected set of compute engines, mapping a computation to this architecture is trivial. However, when considering the underlying topology, it is possible to make decisions that result in varying performance by making poor choices when assigning nodes. Regardless of the choices made though, the CGRA remains usable as long as as there are enough compute engines remaining in the system to be usable, regardless of their orientation relative to one another. It should also be clear that there is a greater opportunity for heterogeneity in the selection of compute engines, as replacing a given compute engine with one with different features doesn't invalidate potential allocations.

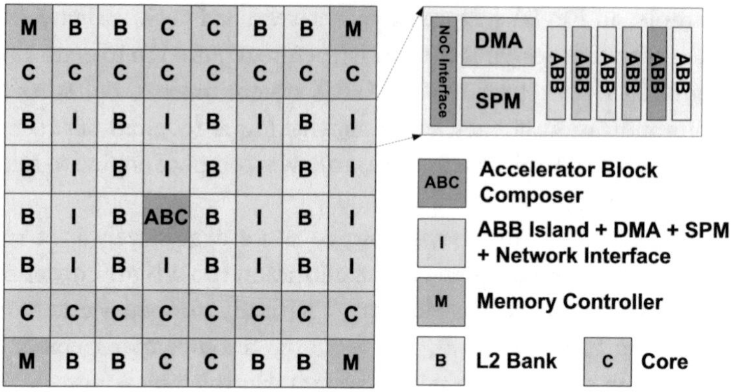

Figure 4.6: A diagram of a CHARM processor, featuring cores and ABBs bundled into islands. Taken from [26].

4.4.3 CHARM

An example CGRA architecture is CHARM [26], shown in Figure 4.6, which focuses on virtualization and rapid scheduling. CHARM introduces a hardware resource management mechanism called an accelerator block composer (ABC), which manages work done by a series of small compute engines called accelerator building blocks (ABBs) that are distributed throughout the processor in a series of islands. There is a single ABC on a chip, and this device has control over a large number of ABBs which are distributed throughout the processor. The ABC acts as a gateway through which a processor can interact with accelerators. Internal to each island, there is a DMA that choreographs data transfer in and out of the island, an amount of scratchpad memory (SPM) for use as ABB buffers, an internal network to facilitate intra-island communication, and a network interface to enable inter-island communication. An island is illustrated in Figure 4.7.

A conventional core invokes an accelerator by writing a configuration to normal shared memory that describes a communicating acyclic graph of ABBs. The ABC then schedules this graph of ABBs among free resources with the objective of maximizing performance for the newly instantiated accelerator, and assigns to each involved ABB a portion of work to perform. To further boost performance, the ABC continues to schedule additional instances of the accelerator until it either runs out of resources, or runs out of work to assign. When all the work is done, the ABC signals to the calling core that work has been completed.

Using the ABC, the hardware scheduling mechanism CHARM is able to recruit a large volume of compute resources to participate in any task that has ample parallelism.

4.4.4 USING COMPOSABLE ACCELERATORS

As with an FPGA, a CGRA platform can be targeted by a compiler. If the compute engines in a CGRA are comparable to the primitive operators used in programming languages (i.e., float-

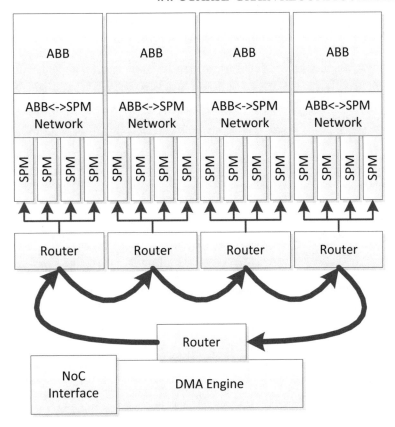

Figure 4.7: An island of ABBs and associated internal structures. This example island features a wide unidirectional ring for internal connectivity, and uses the normal processor interconnect for inter-island communication. Taken from [26].

ing point operations, integer arithmetic, etc.) the mapping of a program region to a CGRA is relatively straightforward. In a practical sense, the process of mapping a program to a CGRA is identical to that of mapping custom instructions described in Chapter 3.4.4, with the added expectation that the CGRA will likely be used to target computations that operate over large volumes of data.

CHAPTER 5

On-Chip Memory Customization

5.1 INTRODUCTION

On-chip memory provides local memory for computational units, such as general processors and accelerators, for better efficiency to access data. Instead of directly fetching data from off-chip DRAM, on-chip memory can provide both short access latency and high bandwidth to hide DRAM access latency. The data blocks with high locality can be cached in on-chip memory. In this chapter, we will first introduce different types of on-chip memories and then discuss the customization techniques for different types of on-chip memory system designs.

5.1.1 CACHES AND BUFFERS (SCRATCHPADS)

Caches are widely used in modern processors to effectively hide the data access latency and improve effective memory bandwidth via data reuse. The data blocks with higher spatial locality or temporal locality can be kept in caches. For general-purpose processors, caches are usually managed automatically through purely-hardware mechanisms. A cache is primarily composed of two arrays: (1) the tag array and (2) the data array. The tag array is used to check whether a request data block is located in the data array. Associative caches are organized into sets, and blocks are managed within a set under the guidance of a cache replacement policy. This policy determines which blocks should be evicted from the cache to make room for other blocks. The memory access patterns in most applications have good spatial and temporal locality, and therefore, the use of caches is an effective technique to improve performance for general-purpose processors.

For applications with predictable data access patterns, it can be more efficient to explicitly manage the on-chip memory based on the access patterns instead of using general replacement policies like caches. We use the terminology *buffers* in this book to represent this type of on-chip memories. A buffer can be logically treated as a one-dimensional array, which can be managed by hardware or software. *FIFOs*, *LIFOs*, and *random access memories* are common hardware-managed buffers used for the on-chip memory system. *Scratchpads* are buffers that are managed by software. Embedded architectures use scratchpads in conjunction with caches to reduce power consumption. The CELL processor is one example that demonstrates the efficiency of utilizing buffers in synergistic processor elements (SPEs) [105].

The comparison between caches and buffers is discussed as follows.

Programmability

Caches are transparent to software and programmers and serve as near memories for general-purpose cores. In general, they are accessed in the same address space as the main memory. The cache blocks in a cache are automatically managed by the hardware cache replacement policy. Most programmers do not need to worry about data management during software development. Some applications benefit from software-based cache optimization (e.g., tiling), but it is not required for cache use. Therefore, caches require minimal programming efforts.

Buffers also serve as near memories for special-purpose cores and accelerators. Buffers can be treated as a one-dimensional array and are addressed in an independent memory space with respect to the main memory. For the hardware-managed buffers, such as FIFO, LIFO and random access memories, programmers do not need to worry about the underlying management policy. The hardware logic for management is determined at the design time. For scratchpads, the programmer or a well-designed compiler is responsible for data management. This creates a burden for programmers to handle data management. However, it also provides freedom for better buffer management strategies by using software for performance optimization. Therefore, buffers are usually used for architectures, such as accelerator-rich architectures, that have predictable memory access patterns.

Performance

Both caches and buffers are used to hide the access latencies to off-chip DRAM and can provide reasonable performance gain. For applications with complicated memory access patterns or access patterns that cannot be predicted at compile time, caches are better choices. Many applications that run on top of general-purpose cores have this kind of behavior. Buffers are usually used for special-purpose cores and accelerators, which usually run applications or computation kernels with predictable memory access patterns. It is common for an accelerator to request multiple data elements per computation. In order to fetch multiple data elements simultaneously with the same fixed latency, designers usually provide multiple individual buffers to perform banking during the accelerator design time. The throughput of the accelerator can thus be significantly improved. Although the internal banking provided in a cache can interleave simultaneous accesses, it does not guarantee that all of the accesses can be interleaved successfully without conflicts. Therefore, buffers are superior in this case, especially when the access patterns are known in the design time of the accelerator.

Another advantage of a buffer is the predictable access latency. The access latency when using caches is difficult to predict since the cache blocks are managed by most general replacement policies. A cache miss can occur during runtime, which triggers the cache controller to send a request to the memory controller to retrieve a data block. The access latency is unpredictable since the block may be at different levels of the cache hierarchy and there may be contention for memory hierarchy resources. In contrast, buffers can provide absolutely predictable performance since the data blocks are explicitly managed by software or hardware. Buffers can guarantee that critical performance targets are met. With compiler optimizations, buffers can further achieve better data reuse and save next-level memory accesses.

Energy and Area

Buffers are more energy- and area-efficient than caches. This is because a cache utilizes a tag array for finding data in the data array. Therefore, the extra tag array and comparison logic make the area of a cache larger. This also leads to more leakage power consumption. Moreover, a cache consumes more dynamic power since tag comparisons are performed to find a cache block. For an L1 cache, parallel tag comparisons are used to accelerate the process and consume more power. For buffers, no tag comparisons are required and thus they are more energy-efficient than caches [75].

5.1.2 ON-CHIP MEMORY SYSTEM CUSTOMIZATIONS

In this section we give a brief overview of the customization techniques on different categories of on-chip memory. We divide the customization techniques into four categories based on the on-chip memory system: (1) caches for general-purpose processors, (2) buffers for accelerator-based architectures, (3) hybrid caches: providing buffers in caches, and (4) caches with hybrid technologies.

Caches for General-Purpose Processors

State-of-the-art multiprocessors provide a large fraction of the chip area for multilevel on-chip caches. Below the L1 caches, the leakage energy consumption significantly dominates the dynamic energy consumption. The main reason is because the leakage energy is linearly proportional to the number of on-chip transistors. Also, the accesses to the level 2 (L2) cache and the last-level cache (LLC) are less infrequent and thus the dynamic energy consumption is less important. Therefore, the major research efforts of the memory system customization of the CPU have been through reconfiguration to power on/off partial caches to reduce leakage energy. We will discuss some early work, such as selective cache ways [1], DRI-i cache [107], and cache decay [77], in Section 5.2.

Buffers for Accelerator-Rich Architectures

In an accelerator-based architecture, buffers are commonly used as near memories for accelerators. An accelerator needs to fetch multiple input data elements simultaneously with predictable or even fixed latency to maximize its performance. Therefore, buffers are more suitable than caches in such architectures. The accelerator store [89] and CHARM [26] are accelerator-rich architectures that utilize buffers. We will discuss the designs and mechanisms in Section 5.3.

Hybrid Caches: Providing Buffers in Caches

In a customized CSoC, general-purpose cores and a sea of accelerators will be placed on the same chip to satisfy challenging computation tasks. Large multilevel caches are widely used in general-purpose CPUs. Therefore, the study of how to leverage the memory resource of current multilevel caches for accelerators or application-specific processors, such as the CELL processor [105], is an important topic.

The memory resource provided caches can be used as software-managed scratchpads, hardware-managed FIFOs, or random access memories. The scratchpads can improve performance by explicitly managing memory accesses through optimized software with known memory access patterns. Moreover, the energy consumption can be reduced since the tag comparison and parallel data fetching in set-associative caches are avoided. Scratchpads are deployed in the CELL processor [105] and many embedded processors. The idea to provide scratchpads in caches has been explored since the early 2000s [20, 32, 43, 80, 96, 113].

For accelerator-rich architectures, the on-chip memory system design for accelerators is important. Recent work investigates [29, 53, 89] how on-chip memory resources can be shared and utilized as hardware-managed random access memories, FIFOs and software-managed scratchpads. The buffer-integrated-cache (BiC) [53] and buffer-in-NUCA (BiN) [29] demonstrate the effectiveness of providing buffers from the shared L2 cache for accelerators. These techniques will be discussed in Section 5.4.

Caches with Hybrid Technologies

At the end of this chapter, we will discuss the customization strategies for caches with hybrid technologies. Non-volatile memories (NVMs) are emerging technologies which are alternatives for current SRAM and DRAM. NVMs have both low leakage power and high density, and therefore provide us opportunities to create larger on-chip memory with very low leakage consumption. However, endurance, i.e., cell life time, and high dynamic write energy are the drawbacks of NVMs. To alleviate the negative impacts on endurance and high write energy, hybrid caches with both SRAM and NVM are proposed [12, 121, 132]. We will give a brief introduction to hybrid caches and discuss the corresponding customization techniques [16, 17, 76, 131, 132] in Section 5.5. Some of the customization techniques for the hybrid caches also share the ideas of SRAM caches, as discussed in Section 5.2.

5.2   CPU CACHE CUSTOMIZATIONS

In modern general-purpose processors, multilevel caches serve as on-chip memories to hide DRAM access latency. Level 1 (L1) caches are closest to the processor cores and thus can provide superior spatial locality for the data elements in the same cache line. L1 caches are generally private to the cores. Typically, there is one L1 instruction (L1-I) cache and one L1 data (L1-D) cache for each core. Level 2 (L2) caches can be shared by a number of cores and can provide data for L1-I caches and L1-D caches on a miss. Last-level caches (LLC) are usually large, shared by all CPU cores, and can provide data for L2 caches on an L2 miss.

A multilevel cache hierarchy can occupy more than 50% of the total die area in a modern processor [114]. Therefore, leakage energy accounts for a significant portion of energy consumption. In this section we discuss some early but important work in customization strategies for adapting cache sizes to cache demand at runtime to reduce leakage energy. These strategies can be categorized based on the level of granularity to partially power on/off a cache. For example, selective cache ways [1] and the resizable i-cache [107, 134] provide a coarse-grain level of reconfiguration, such as way-based and set-based reconfigurations. The cache decay [77] idea and the drowsy cache [56] provide fine-grain control of cache blocks to reduce leakage energy. Most of these ideas were evaluated on the cache hierarchy with only one or two levels, i.e., L1 cache only or two-level cache. However, the ideas can be adapted to multilevel caches, especially in the LLC for leakage reduction. Although these works are summarized well in a recent Synthesis lecture [78], the optimization techniques used in these works can be also extended well to a cache with disparate memory technologies, e.g., a cache with both SRAM and spin-torque transfer magnetoresistive RAM (STT-RAM) cells [132]. Therefore, we first discuss the techniques developed for the SRAM caches in this section and then discuss more sophisticated techniques that can be applied to the hybrid-technology caches in Section 5.5.

5.2.1   COARSE-GRAIN CUSTOMIZATION STRATEGIES

The selective cache ways architecture [1] can disable a subset of the ways in a set-associative cache when the cache demand is low. The on-demand resource allocation can dynamically reduce the leakage energy dissipation based on application need. It is a good example of the coarse-grain customization strategy since the proposed architecture is configured at the way granularity.

Figure 5.1 shows the proposed architecture of selective cache ways. First, the decision logic and gating hardware are required to disable cache ways. The gating hardware uses *gated-Vdd* technique implemented in the circuit level [107]. Second, the *cache way select register* (*CWSR*) is used to signal the cache controller to enable/disable particular cache ways. The CWSR is software-visible. Additional ISA support is required to read and write the CWSR. Third, the author suggested that the cache demand can be estimated dynamically through software-based techniques, such as profiling tools and periodic performance counter sampling. Finally, a proper coherence state needs to be maintained when the cache way is disabled. The results show that a significant amount of leakage consumption can be reduced with little performance loss. For example, a 64KB 4-way L1-D cache with 1MB L2 cache sees a 30% average cache energy savings with less than 4% performance degradation [1].

While the authors in [134] proposed the resizable i-cache (DRI i-cache) to dynamically resize and adapt to the application behaviors, the underlying circuit-level technique to realize DRI i-cache is *gated-Vdd* [107], which utilizes the stacking effect of self-reverse-biasing series-

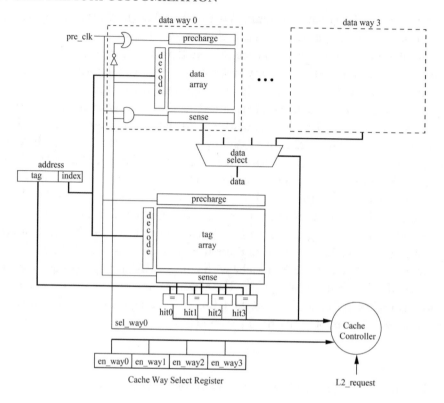

Figure 5.1: A 4-way set-associative cache using selective cache ways. Taken from [1].

connected transistors [135]. With the support of the circuit-level technique, a cache can be configured at a certain granularity dynamically.

Figure 5.2 demonstrates the architectural design of the DRI i-cache. The DRI i-cache can dynamically record the miss rate within a fixed-length interval. This is achieved through a *miss counter*, a *miss-bound*, and the *end of interval flag*. The *miss counter* records the number of misses in an interval; this is used to measure the cache demand of the DRI i-cache. The *miss-bound* is a value preset to be the upper bound of misses. If the miss counter is larger than the miss bound, the DRI i-cache is then downsized. Otherwise, the DRI i-cache is upsized. The cache size can be changed by a factor of two by alternating the number of bits used in the index. To prevent the cache from thrashing and downsizing to a prohibitively small size, a *size-bound* is provided to specify the minimum cache size. A DRI i-cache can reduce the overall energy-delay product by 62% while increasing the execution time by at most 4%.

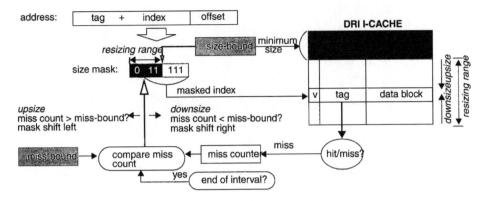

Figure 5.2: Resizable i-cache (DRI i-cache). Taken from [107].

5.2.2 FINE-GRAIN CUSTOMIZATION STRATEGIES

For fine-grain strategies, we use the cache decay idea [77] as an example of how block-level manipulation can save energy. Figure 5.3 shows the lifespan of a data block from when it is brought into the cache until it is evicted [77]. The authors in [77] discovered that a cache line would only be frequently accessed when the cache line is first brought in. After that, the data in the cache line would not be reused anymore until the new data item replaces the old data in the cache line. The time period between the last access of the cache line and the cache replacement time point is called the *dead time*. Figure 5.4 shows the percentage of the dead time of cached data from profiling. On average, the dead time is 65% for integer benchmarks and 80% for floating point benchmarks. Based on observation, if the cache lines during dead time can be powered off, the leakage energy can be significantly reduced. The authors proposed *cache decay*, which is a time-based leakage control technique, to automatically power off the cache lines during the dead time [77]. The cache decay strategy is performed at the fine-grain cache line granularity.

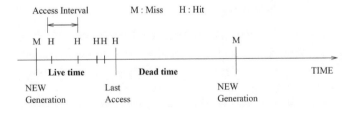

Figure 5.3: Cache generations in a reference stream. Taken from [77].

The key idea is to use a counter, which is incremented periodically, to measure if a cache line has not been accessed for a long period of time. If the cache line is accessed, the counter

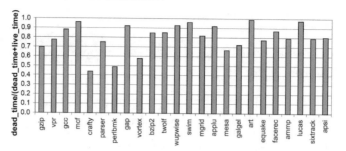

Figure 5.4: Fraction of time cached data are "dead." Taken from [77].

would be reset. If the counter saturates, it means that the cache line is unlikely to be accessed and thus can be powered off. However, the decay interval can be at the range of tens of thousands of cycles. The counter design can thus be impractical since it requires a large number of bits.

Therefore, the authors proposed a hierarchical counter design where a single global counter is used to provide the ticks for smaller cache-line counters, as shown in Figure 5.5. The local two-bit counter is reset when the corresponding cache line is accessed. To avoid the possibility of a burst of write-backs when the global tick signal is generated, the tick signal is cascaded from one local counter to the next with a one-cycle latency. The proposed cache decay idea can reduce the L1 leakage energy by 5x for the SPEC CPU2000 with negligible performance degradation [77].

In conclusion, the fine-grain strategies have higher flexibility and can have better leakage reduction within a performance degradation limit compared to coarse-grain strategies. However, the control logic design and circuit-level design may introduce overhead in area consumption and routing resources. This further complicates the design to implement fine-grain strategies. These trade-offs need to be considered by designers when choosing a cache customization strategy.

5.3   BUFFERS FOR ACCELERATOR-RICH ARCHITECTURES

As discussed in Section 5.1.1, buffers have two major advantages over CPU caches: (1) predictable access latency and (2) flexible banking. Therefore, they are commonly used for application-specific accelerators in today's SoC designs. Recent research focuses on how to design an on-chip memory system with buffers that can be shared for accelerator-rich architectures [29, 53, 89]. Other work has focused on how to optimize the buffers and interconnect microarchitectures based on the access patterns of an accelerator [21, 33]. In this section we first use the accelerator store [89] to describe how current accelerator-rich architectures design shared buffer systems. The techniques to provide buffers within a CMP's caches [29, 53] will be discussed in Section 5.4.2. Next, we discuss the customization technique of buffers inside an accelerator based on its the specific memory access patterns. We use the stencil computation as an example application [33].

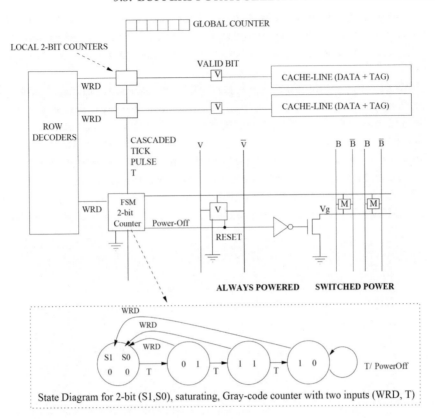

Figure 5.5: Cache decay: hierarchical counters. Taken from [77].

5.3.1   SHARED BUFFER SYSTEM DESIGN FOR ACCELERATORS

The accelerator store (AS) [89] tries to provide a shared memory framework for accelerator-rich systems, which contain tens or hundreds of accelerators. When the number of accelerators increases, the amount of SRAM memory required by accelerators can dominate the area for such systems. Therefore, memory resource sharing is required.

The AS uses *handles* to create shared memory for accelerators. A handle represents a shared memory store in the AS, which is similar to the way a file handle represents a file in the C programming language. Each accelerator can contain several shared memories in the AS. When a shared memory is created in the AS, the corresponding handle ID (HID) is returned. The AS passes the HID to one or many accelerators. The accelerators can then exchange data using the assigned HID.

Figure 5.6 shows the design of the accelerator store. The handle table maintains each shared memory in the AS by recording the handles for each active shared memory. The system software running on the general-purpose CPU (GPCPU) can add or remove handles. The requests to a shared memory are transmitted over channels. Each channel can carry one shared memory access per cycle. The priority tables can arbitrate the shared memory accesses to resolve the contentions for channels.

The AS can support the following types of buffers, which are managed through hardware logic. First, it supports random access memory. Accelerators can read or write data at any location in the shared memory. Second, it supports FIFOs. Accelerators can put data in or get data out in FIFO order. FIFOs can be used for inter- or intra-accelerator queues. Third, it supports FIFOs with random access capability. In this type of buffer, the reads and writes do not add or remove values. This is useful for operations that stream in data in FIFO order, but the data can be used out of order. The AS is evaluated for two applications and significant system area reductions (30%) can be achieved from buffer sharing with small overheads (2% performance, 0%–8% energy) [89].

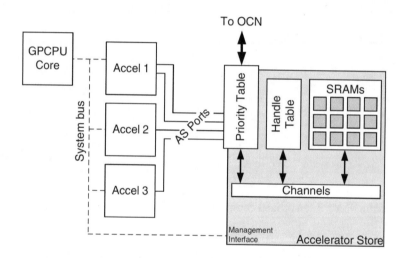

Figure 5.6: The accelerator store design. Taken from [89].

5.3.2   CUSTOMIZATION OF BUFFERS INSIDE AN ACCELERATOR

Figure 5.7 demonstrates the memory access patterns of a classic five-point stencil computation. In each iteration, we need to update the values of each cell. To achieve the update, each cell needs to fetch its four neighbors' values before computation. To design an accelerator with maximum performance, we expect that the five data elements can be fetched simultaneously and arrive at the accelerator at the same cycle for computation. If buffers are used, the accelerator can leverage the benefits of the fixed access latencies of the five data elements and access five elements from five buffers without conflicts. The accelerator can thus update one cell per cycle, which means the initiation interval is one (II=1). If caches are used instead, the indeterminate access latencies to fetch five elements may severely degrade the accelerator performance.

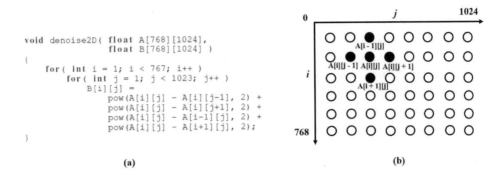

```
void denoise2D( float A[768][1024],
                float B[768][1024] )
{
    for( int i = 1; i < 767; i++ )
        for( int j = 1; j < 1023; j++ )
            B[i][j] =
                pow(A[i][j] - A[i][j-1], 2) +
                pow(A[i][j] - A[i][j+1], 2) +
                pow(A[i][j] - A[i-1][j], 2) +
                pow(A[i][j] - A[i+1][j], 2);
}
```

(a)   (b)

Figure 5.7: A classic five-point stencil computation. Reproduced from [33].

In [33], the authors provide an optimal microarchitecture that minimizes the number of buffers and the size of each buffer for stencil computation. The details are covered in Chapter 6 and in [33].

5.4 PROVIDING BUFFERS IN CACHES FOR CPUS AND ACCELERATORS

In Section 5.3 we briefly discussed on-chip shared buffer systems. These buffers are independent of the CPU caches, which are rich on-chip memory resources in a chip multiprocessor. In this section we will discuss the strategies to provide buffers by leveraging the abundant on-chip cache resources. The required buffer resources are provided from CPU caches through architectural support and can be used for CPUs or accelerators. In this section, we will use the term "*hybrid caches*" to refer to an on-chip memory system mixed with both caches and buffers.

In this section we will discuss two strategies to utilize CPU cache resources. First, the cache resource can be used as software-managed scratchpads [20, 32, 113]. We illustrate the idea by discussing an early study, column caching [113] and a more sophisticated recent study, adaptive hybrid cache (AH-cache) [32]. Second, for accelerator-rich architectures, the cache resource can be utilized as buffers or FIFOs for accelerators. We will discuss the buffer-integrated-cache (BiC) [53] and buffer-in-NUCA (BiN) [29] to show how CPUs and accelerators can simultaneously share the cache resource in an efficient way. Furthermore, the required architectural support for hybrid cache architectures is discussed in both strategies. Another important issue for hybrid caches is to prevent performance loss when resource contention between caches and buffers occurs.

5.4.1   PROVIDING SOFTWARE-MANAGED SCRATCHPADS FOR CPUS

Coarse-Grained Method

The authors in [113] proposed a cache architecture, called reconfigurable caches, to dynamically divide SRAM cache arrays into multiple partitions that can be used for different application activities. First, the different cache partitions can be used for hardware optimizations using lookup tables or buffers. For example, the value prediction, memoization, and instruction reuse techniques that have been studied require lookup tables. Second, these partitions can be used to store software and hardware prefetched data. Third, they can be used as scratchpads, managed by compilers or user applications.

For a set-associative cache, reconfigurable caches provide partitions through cache ways, as shown in Figure 5.8. The granularity is at the cache way, which is similar to column caching. The changes to the conventional cache are (1) multiple input and output paths and (2) a cache status register. An N-way reconfigurable cache can support up to N input addresses and provide N output data elements. A cache status register is used to track the number and sizes of the partitions. This register would be preserved when context switches occur, like any other control register in a processor.

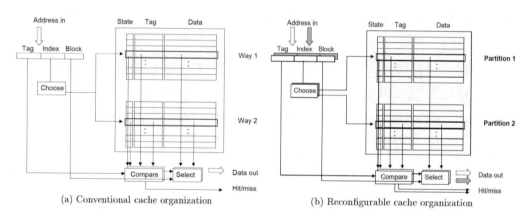

Figure 5.8: Associativity-based partitioning for reconfigurable caches. Taken from [113].

For a direct-mapped cache, the authors use the overlapped wide-tag partitioning scheme, as shown in Figure 5.9. The additional tag bits are used to indicate the partitions. The number of partitions is limited to be powers of two for simpler decoding. The results show that IPC improvements range from 1.04x to 1.20x for eight media processing benchmarks.

Fine-Grained Method: Adaptive Hybrid Cache (AH-Cache)

As discussed in Section 5.4.1, the authors proposed the reconfigurable caches to enable the cache to be dynamically partitioned for specific use. However, this hybrid cache design partitions the

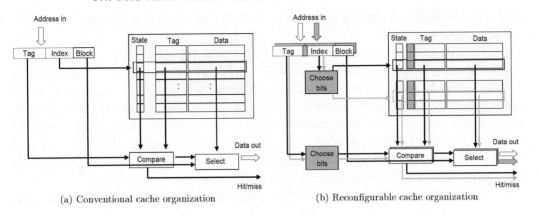

(a) Conventional cache organization   (b) Reconfigurable cache organization

Figure 5.9: Overlapped wide-tag partitioning for reconfigurable caches. Taken from [113].

cache and scratchpads without adapting to the run time cache behavior. Since cache sets are not uniformly utilized [110], this uniform mapping of SPM blocks onto cache blocks may create hot cache sets at run time, which will increase the conflict miss rate and degrade the performance. Figure 5.10 shows the cache set utilization statistics for a hybrid cache. Each column represents a set in the cache, while each row represents one million cycles of time. A darker point means a hotter cache set. We see that the cache set utilizations vary a lot for different applications. For a given application, the utilization still varies over time. A good adaption technique is required for the development of a hybrid cache.

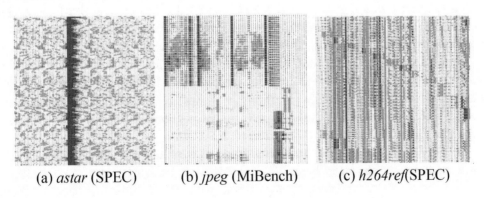

(a) *astar* (SPEC)   (b) *jpeg* (MiBench)   (c) *h264ref*(SPEC)

Figure 5.10: Non-uniformed cache sets utilization in a hybrid cache. Taken from [32].

Two challenges are encountered for an efficient hybrid cache design. The first challenge is how to balance cache set utilization when SPMs are allocated in the cache. The second challenge is how to efficiently find the blocks in SPMs when the blocks are frequently remapped to different cache blocks. Therefore, hardware support is required. The software only focuses on the use of a logically continuous SPM.

The authors in [32] proposed the *adaptive hybrid cache* (AH-Cache) to address these challenges. First, the lookup operation of the SPM location is hidden in the execution (EX) stage of the pipeline of the processor, and a clean software interface is provided as a non-adaptive hybrid cache. Second, a victim tag buffer, similar to the missing tag [137], is used to assess the cache set utilization by sharing the tag array, resulting in no storage overhead. Third, an adaptive mapping scheme is proposed for fast adaptation to the cache behavior without the circular bouncing effect. The circular bouncing effect means that the allocated SPM blocks keep bouncing between several hot cache sets, which incurs energy and performance overheads.

Figure 5.11 shows an example of SPM management in AH-Cache. The system software is provided with two system APIs to specify the scratchpad base address and size. As shown in Figure 5.11(b), *spm_pos* sets the *SPM base address register* as the address of the first element of array *amplitude*, and *spm_size* sets the SPM size register as the size of the array *amplitude* and *state*. Note that these system APIs do not impact the ISA since they use regular instructions for register value assignment.

As shown in Figure 5.11(d)(e), the partition between cache and SPM in AH-Cache is at a cache-block granularity. If the requested SPM size is not a multiple of a cache block, it will be increased to the next block-sized multiple. The mapping information of SPM blocks onto the cache blocks is stored into an *SPM mapping lookup table* (SMLT). The number of entries in SMLT is the maximum number of cache blocks that can be configured as SPM. Since AH-Cache must hold at least one cache block for each cache set to maintain the cache functionality, the maximum SPM size on a M-way N-set set-associative cache is (M-1)*N blocks. In each SMLT entry, there are (1) a valid bit indicating whether this SPM block falls into the real SPM space, since the requested SPM size may be smaller than the maximum SPM size; and (2) a set index and a way index which locate the cache block upon which the SPM block is mapped.

AH-Cache needs an additional step in order to use the low-order bits of the virtual address to look up the SMLT. This further increases the pipeline critical path. To solve this problem, inspired by the zero-cycle load idea [3], the address checking and SMLT lookup are performed in parallel with the virtual address calculation of the memory operation in a pipelined architecture, as shown in Figure 5.12.

In AH-Cache, the low-demand sets are used to accommodate more SPM blocks than the high-demand sets, as shown in Figure 5.11(e). Miss rate cannot be used to recognize a high-demand cache set, since for streaming applications with little locality or applications hopelessly thrashing the cache, even if the miss rate is high, there is little benefit in increasing the cache blocks. AH-Cache uses a victim tag buffer (VTB) to capture the demand of each set. This is similar to the miss tag introduced in [137], but with no memory overhead. The details of the VTB management can be found it [32].

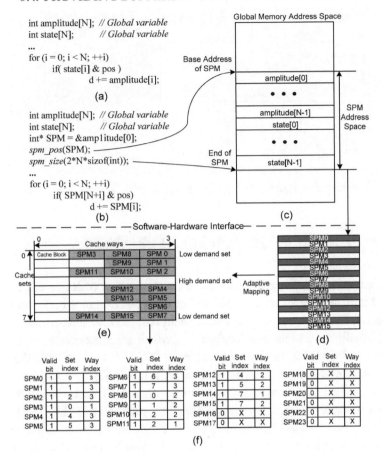

Figure 5.11: (a) Original code, (b) transformed code for AH-Cache, (c) memory space view of SPM in AH-Cache, (d) SPM blocks, (e) SPM mapping in AH-cache, (f) SPM mapping lookup table (SMLT). Taken from [32].

We call the blocks bouncing around different sets "floating blocks." AH-Cache uses a *floating block holder* queue and a specific *reinsertion bit table* to handle the circular bouncing problem and perform adaptive block mapping. Readers can refer to [32] for more details.

Overall, the AH-Cache can reduce the cache miss rates by up to 52% and provide around 20% energy-delay product gain over prior work.

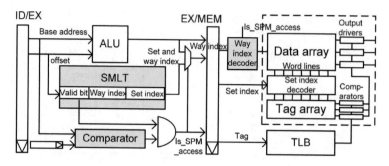

Figure 5.12: SPM mapping lookup and access in AH-Cache. Taken from [32].

5.4.2 PROVIDING BUFFERS FOR ACCELERATORS

Buffer-Integrated-Cache

In an embedded processor, the accelerator blocks such as video codec, image processors, crypto, and network interface controllers are used to improve performance and energy efficiency. For these accelerators, SRAM blocks are used as scratchpads, FIFOs, lookup tables, etc. These memory blocks are usually implemented as local buffers, which are distinct from caches, for each accelerator, as shown in Figure 5.13(a). The researchers in BiC [53] try to explore the architecture that allows buffers and a cache to reside in the same SRAM block. Figure 5.13(b) shows the proposed BiC architecture. The buffers are allocated from the shared L2 cache in the SoC. The general-purpose cores can share the memory resource with accelerators in BiC. The shared BiC cannot eliminate the need for small local buffers or registers inside an accelerator for high-bandwidth and low-latency accesses. However, it can provide large buffers by leveraging the memory resources of a large shared L2 cache.

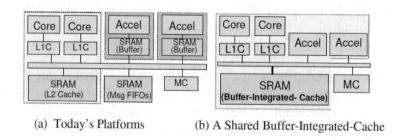

(a) Today's Platforms   (b) A Shared Buffer-Integrated-Cache

Figure 5.13: Sharing buffers in an SoC. Taken from [53].

A buffer in BiC can be identified by a set index and a way index and is allocated at the cache line granularity. Buffer allocation is performed by allocating contiguous cache lines within the same cache way across cache sets. This can avoid the starvation problem if buffers are allocated in the same set. Moreover, disparate buffers must not overlap in the cache lines.

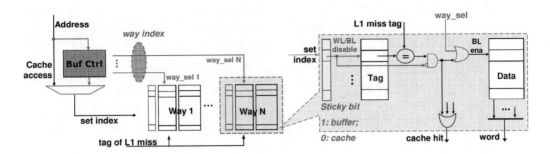

Figure 5.14: BiC implementation. Taken from [53].

Figure 5.14 shows the implementation of BiC. An additional buffer/cache (B/C) sticky bit is added to each cache line. If the line is used by a buffer, the sticky bit is set to 1. If the line is used for cache, it is set to 0. The replacement logic would avoid selecting the cache lines with their B/C sticky bit equal to 1 as replacement candidates. For a buffer operation, the B/C sticky bit can be set to disable tag comparison. Instead, the *way_sel* signal is used to fetch the data in the data array. Therefore, the dynamic power from tag comparison can be eliminated.

BiC can eliminate the need to build a 215KB dedicated SRAM for accelerators with only 3% extra area added to the baseline L2 cache. The cache miss rates only increase by no more than 0.3%.

Buffer-in-NUCA

The buffer-in-NUCA architecture (BiN) [29] further extends the BiC work from allocating buffers in a centralized cache to a non-uniform cache architecture (NUCA) with distributed cache banks. In NUCA there are multiple physically distributed memory banks in the L2 cache or LLC. Figure 5.15(a) shows the architectural overview of BiN. BiN further extends the accelerator-rich architecture [28] discussed in Section 4.2. ABM stands for the accelerator and BiN manager, which is a centralized controller to manage the accelerator and BiN resource. The interactions between the core, ABM, accelerators, and L2 cache banks are described in Figure 5.15(b).

In BiN, the authors aim to solve the following two problems: (1) how to dynamically assign buffer sizes to accelerators that can best utilize buffers to reduce the off-chip bandwidth demand, and (2) how to limit the buffer fragmentation during allocation.

In general, the off-chip memory bandwidth demand can be reduced by increasing the sizes of accelerator buffers, which is discussed in [42]. This is because longer data reuse distance can be covered with larger buffers. The trade-offs between *buffer sizes* and *bandwidth demands* can be depicted as a curve, which is called *BB-Curve*. In BiN ABM can collect the buffer requests in

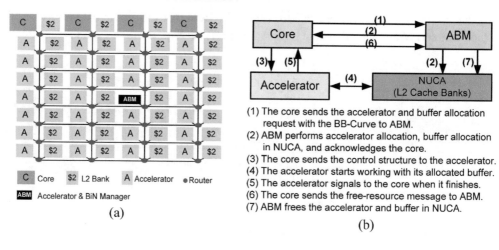

(1) The core sends the accelerator and buffer allocation request with the BB-Curve to ABM.
(2) ABM performs accelerator allocation, buffer allocation in NUCA, and acknowledges the core.
(3) The core sends the control structure to the accelerator.
(4) The accelerator starts working with its allocated buffer.
(5) The accelerator signals to the core when it finishes.
(6) The core sends the free-resource message to ABM.
(7) ABM frees the accelerator and buffer in NUCA.

(a)

(b)

Figure 5.15: BiN: architecture overview. Taken from [29].

a short fixed-time interval and then perform the global allocation for the collected requests. An optimal algorithm that can dynamically allocate the buffers is proposed to guarantee the short-time optimality. The details of the algorithm can be found in [29].

In order to simplify the buffer allocation and the location decode of buffer accesses, prior work allocates buffers in a physically contiguous fashion. This may introduce fragmentation, especially when many accelerators request buffer resources. In BiN, the authors propose paged buffer allocation, borrowing the idea from virtual memory, to provide flexible buffer allocation at a page granularity. The page size of each buffer can be different and is able to adapt to the buffer size required by an accelerator. Figure 5.16 demonstrates an example of buffer allocation on three accelerators. The buffer allocator in ABM first selects the nearby L2 banks for allocation. To reduce the page fragments, BiN allows the last page of a buffer to be smaller than the other pages of this buffer. This does not affect the page table lookup. Therefore, the maximum page fragment for any buffer is smaller than the minimum page size.

The two major hardware components to support BiN are (1) the buffer allocator module in ABM, and (2) the buffer page table and address generation logic of an accelerator. Figure 5.17 shows the design of the buffer allocator module in the ABM. For each L2 buffer bank, the buffer allocation status of each cache way needs to be recorded. For an L2 cache with N ways, a $(N - 1)$-entry table is used to keep track of the allocation status for each bank. At most, $N - 1$ ways can be allocated for buffers to prevent starvation. The maximum number of buffer requests that can be handled in the fixed-time interval is set to a given number (eight in this example). Therefore, eight SRAM tables are reused to record the BB-Curve points.

For each accelerator, a local buffer page table is required to perform address translation, as shown in Figure 5.18. For a 2MB L2 cache with 32-bank and 64-byte cache line, a 5-bit cache

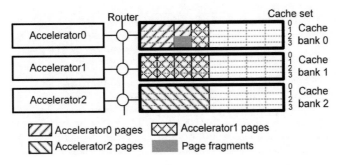

Figure 5.16: BiN: an example of the paged buffer allocation. Taken from [29].

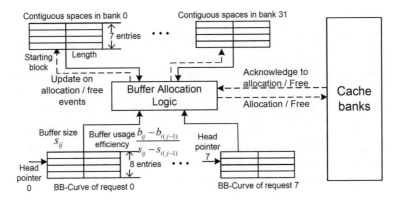

Figure 5.17: BiN: the buffer allocator module in ABM. Taken from [29].

bank ID and a 10-bit cache block ID are required. Compared to BiC, BiN can further improve performance and energy by 35% and 29%, respectively.

5.5   CACHES WITH DISPARATE MEMORY TECHNOLOGIES

Conventionally, SRAM has been used for both hardware logic and on-chip multilevel caches. However, the traditional SRAM-based on-chip cache has become a bottleneck for an energy-efficient design due to its high leakage power. Designers have turned their attention toward emerging non-volatile memories, such as the spin-torque transfer magnetoresistive RAM (STT-RAM) and phase change RAM (PRAM), to build future memory systems. Power, performance, and density characteristics of the new memory technologies differ dramatically compared to SRAM, and thus they enlarge the landscape of memory design.

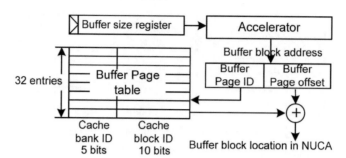

Figure 5.18: BiN: the block address generation with the page table in an accelerator. Taken from [29].

Table 5.1 shows a brief comparison of SRAM, STT-RAM, and PRAM technologies. The exact access time and dynamic power depend on the cache size and the peripheral circuit implementation. In sum, SRAM suffers from high leakage and low density while providing great endurance, i.e., cell lifetime, while STT-RAM and PRAM provide high density and low leakage at the cost of weak endurance. Moreover, STT-RAM outperforms PRAM in terms of endurance, access time, and dynamic power, while PRAM has higher density. Based on endurance, STT-RAM is more suitable for on-chip last-level cache design due to its higher endurance [14, 16, 17, 47, 76, 121, 131, 132], while PRAM is promising as an alternative for DRAM in the main memory design due to its higher density [86]. In this section we focus on the discussion of on-chip memory.

Table 5.1: Comparison among SRAM, STT-RAM, and PRAM. Taken from [16].

	SRAM	STT-RAM	PRAM
Density	1X	4X	16X
Read time	Very fast	Fast	Slow
Write time	Very fast	Slow	Very slow
Read power	Low	Low	Medium
Write power	Low	High	High
Leakage power	High	Low	Low
Endurance	10^{16}	4×10^{12}[14]	10^9

In this section we use the term "hybrid cache" to refer to a cache that uses disparate memory technologies. A hybrid cache can leverage advantages from both SRAM and NVM while hiding their disadvantages. In general, a hybrid cache provides a larger cache size than that of the conventional SRAM-based cache by using higher density NVM cells. Moreover, the leakage consumption is much smaller in a hybrid cache. A hybrid cache can utilize its SRAM cells to hide the drawbacks of low endurance and high dynamic write energy in NVM cells. With the benefits of denser and near-zero leakage NVM cells, a hybrid cache is best suited as the LLC.

Hybrid cache architectures were first proposed in [121, 132]. In [132] the authors try to explore the performance and energy of different types of hybrid cache architectures. The exploration includes disparate NVM technologies, different multilevel configurations, and 2D/3D hybrid caches, as shown in Figure 5.19. In this section, we focus on hybrid cache designs similar to the region-based hybrid cache architecture (RHCA), which considers disparate memory technologies in the same cache level. Readers who are interested in the other design styles can refer to [132] for more details.

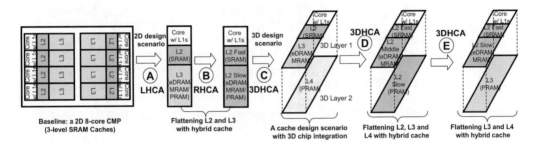

Figure 5.19: Exploration of different hybrid cache architecture configurations. Taken from [132].

Figure 5.20 shows an example of a hybrid cache design presented in [16]. First, a hybrid cache contains data arrays with disparate technologies. Second, the accesses to the tag array and data array are done sequentially (i.e., the data array will be accessed after the tag array). Such a serialized tag/data array access has already been widely adopted in a modern low-level large-scale cache for dynamic energy reduction. Third, the tag array is fully implemented with SRAM cells to avoid long access latencies if NVM are applied. Fourth, a hybrid cache can be partitioned based on the cache way granularity [16, 17, 76, 131, 132]. The NVM region contains more cache ways, i.e., cache capacity, than those of the SRAM region since NVM is denser. Finally, for performance consideration, a hybrid cache with disparate technologies is usually deployed at an L2 cache or a LLC, but not at an L1 cache. This is because the read and write access latency for an NVM block is usually longer than that of a SRAM block, which leads to significant performance overhead if NVM is applied at an L1 cache. However, the access latency of a hybrid cache in an L2 cache or an LLC can be hidden by the SRAM-based L1 cache. For example, when a miss occurs in an STT-RAM line in LLC, a request would be issued to the memory controller to fetch the data back. The fetched data can be directly forwarded to upper-level caches (L1 or L2) without waiting for the whole write process on the STT-RAM block to complete.

However, the baseline hybrid cache can still suffer from the low endurance and high dynamic write energy arising from NVM cells. In this section we will discuss the customization strategies to leverage the NVM benefits while hiding the drawbacks of NVM cells. Similar to the

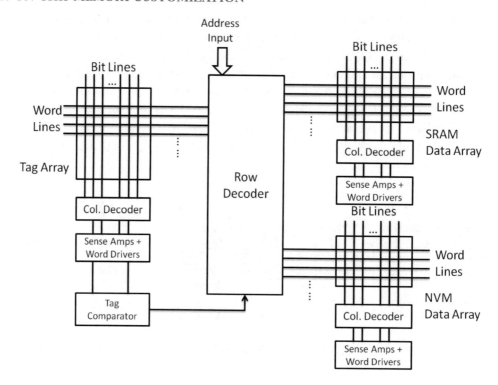

Figure 5.20: Hybrid cache with disparate memory technologies. Taken from [16].

customization techniques discussed in Section 5.2, the customization techniques used in a hybrid cache can be summarized into coarse-grain and fine-grain techniques as well. First, we discussed the coarse-grain techniques, such as the selective cache ways [1] and DRI i-caches [107] in Section 5.2. Similarly, more sophisticated dynamic reconfiguration strategies can be performed at a coarse-grain level for a hybrid cache to reduce leakage [16]. Second, for the fine-grain techniques, we discussed the cache decay idea [77] used for SRAM caches in Section 5.2. However, the hybrid caches with disparate memory technologies require more sophisticated techniques to handle both the SRAM and NVM cache blocks. For example, techniques such as adaptive block placement and block migration between SRAM and NVM cache ways are proposed to hide two drawbacks of NVM blocks: the high dynamic write energy and the low endurance [16, 17, 76, 131, 132]. Note that the dynamic reconfiguration techniques are orthogonal to the fine-grain block placement and migration techniques. Therefore, the coarse-grain technique and the fine-grain techniques can be applied together to optimize performance, energy, and endurance simultaneously.

5.5.1   COARSE-GRAIN CUSTOMIZATION STRATEGIES

Based on the research proposed in [16, 17, 76, 131, 132], leakage consumption still accounts for a significant percentage of the total energy consumption (> 30%) in a hybrid cache even if STT-RAM cells are deployed. Therefore, researchers have explored dynamic reconfiguration techniques for hybrid caches to further reduce leakage. In Section 5.2 we introduced several reconfiguration techniques for SRAM-based caches, such as selective cache ways [1], Gated-Vdd [107], and cache decays [77]. In this section we will introduce the dynamically reconfigurable hybrid cache (RHC) [16], which provides effective methods to reconfigure a hybrid cache.

The RHC architecture shown in Figure 5.20 can be dynamically reconfigured at the way granularity based on the cache demand. Figure 5.21 illustrates the power-gating design adopted in RHC to perform dynamic reconfiguration. A centralized power management unit (PMU) is introduced to send sleep/wakeup signals to power on/off each SRAM or NVM way. The power-gating circuits of each way in SRAM tag/data arrays are implemented with NMOS sleep transistors to minimize the leakage. In this design the stacking effect of three NMOS transistors from the bitline to GND, substantially reduces leakage [107]. In RHC, the SRAM cells in the same cache way will be connected to a shared virtual GND while the virtual GNDs among different cache ways are disconnected. This can ensure that the behaviors of cache ways that are powered-on will not be influenced by the powering-off process in other ways.

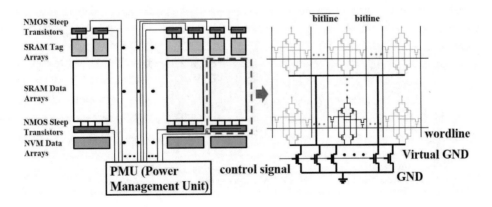

Figure 5.21: PMU and the power-gating design of RHC. Taken from [16].

For the dynamic reconfiguration strategy, the authors addressed the following key questions: (1) how to measure cache demand accurately, (2) how to make power-off decisions without cache thrashing, and (3) how to deal with hybrid data arrays. Figure 5.22 shows the potential hit counter scheme used in [16]. The cache demand is measured by the potential hit counters, which count the potential hits that occur in SRAM or STT-RAM data arrays. A potential hit is a hit that could occur if the cache way was on. This is achieved by comparing the tag of the current access and the tags in the powered-off cache ways. The potential hit idea is similar to the missing tags [32] and victim tags [137]. To address the hybrid data arrays question, RHC provides two potential hit counters for the SRAM data array and STT-RAM data array, respectively. Also, the powered-on and powered-off thresholds are different. When the value of a potential hit counter is over the threshold measured in a given time period, the whole cache way is then powered on/off automatically.

To alleviate the cache thrashing problem that arises from powered-off the whole cache way, the authors in [16] proposed the asymmetric powered-on and powered-off strategy. The powered-on speed is generally faster than that of the powered-off speed and is achieved using the following two rules. First, in the given time period, only one cache way can be powered-off. In contrast, multiple cache ways can be powered-on in the next time period when the measured cache demand is high. Second, to further reduce the impact of the thrashing problem, the powered-off speed is set to be slower. For example, the powered-off action can be triggered only when the low cache demand lasts for a given number of consecutive periods, e.g., 10 periods.

According to [16], the proposed RHC achieves an average 63%, 48%, and 25% energy savings over non-reconfigurable SRAM-based cache, non-reconfigurable hybrid cache, and re-configurable SRAM-based cache, while maintaining the system performance (at most 4% performance overhead) for a wide range of workloads.

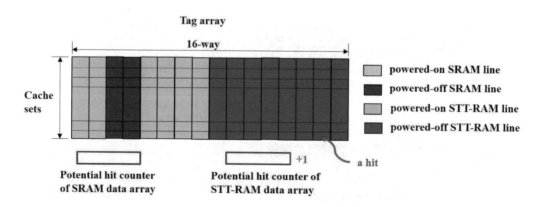

Figure 5.22: Potential hit counters and the tag array.

5.5.2 FINE-GRAIN CUSTOMIZATION STRATEGIES

Dynamic reconfiguration can be used to reduce leakage in a hybrid cache, as discussed in Section 5.5. However, the NVM blocks in a hybrid cache still suffer from two drawbacks: the high dynamic write energy and the low endurance. To alleviate the drawbacks brought by NVM blocks, fine-grain customization strategies, such as block placement and block migration, have been developed for a hybrid cache. We summarize the proposed strategies into three categories: (1) the dynamic migration schemes [76, 132], (2) the combined SW/HW scheme [17], and (3) the dynamic placement and migration scheme [131].

Dynamic Migration Schemes
The authors in [132] first proposed a dynamic migration scheme for a hybrid cache, as shown in Figure 5.23. The initial purpose in [132] is to migrate the most frequently reused blocks from NVM to SRAM to improve performance. The fast region is composed of one SRAM way, while the slow region is composed of many NVM ways. The authors added three new components to provide architecture support for dynamic migration. First, the *sticky bits* in the fast region are used to monitor whether the block in the SRAM way was reused recently. Second, the *saturation counters* in the slow region record the access frequency, i.e., the "hotness" of a block in the slow region. Once the saturation counter is saturated, the block can be migrated to the SRAM block in the same cache set if the sticky bit of the SRAM block is not set. If the sticky bit is set, migration would not occur. The sticky bit is then cleared and the saturation counter is reset. Third, a swap operation involves reading out two cache blocks from two regions and writing each to the other region. However, the swap operation cannot be completed immediately because of the speed difference between two regions. Therefore, a *swap buffer* is designed to temporarily host the cache blocks. The reader can refer to [132] for the migration policy details.

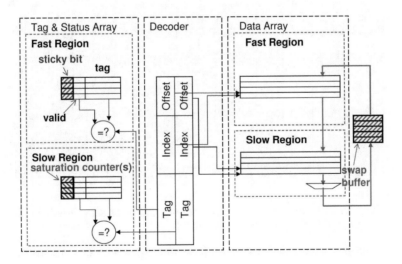

Figure 5.23: Block migration scheme in the region-based hybrid cache architecture (RHCA). Taken from [132].

However, the authors in [132] did not distinguish policies applied between reads and writes to STT-RAM cache lines. The writes to STT-RAM lines are harmful due to their high dynamic write energy and low endurance to STT-RAM cells. The authors in [76] tried to improve the endurance of a hybrid cache by proposing two management policies: intra-set remapping and inter-set remapping. Figure 5.24 shows the hybrid cache architecture proposed in this work.

The intra-set remapping can be divided into two types of migrations: (1) data migration between SRAM and STT-RAM lines, and (2) data migration within STT-RAM lines (in the same set). The data migration between SRAM and STT-RAM lines helps migrate the write-intensive blocks from STT-RAM lines to SRAM lines, while the data migration within STT-RAM lines averages the write intensity for the STT-RAM lines in the same set.

The *line saturation counter* (LSC) is used to monitor the recent write intensity of each line. LSC increments when a write access occurs. When the LSC of a STT-RAM line saturates, the cache controller tries to find a victim with the lowest LSC value within the SRAM lines. The non-uniformity of writes on STT-RAM lines still exists even if the migrations to SRAM lines are performed. The authors used the *wear-level saturation counter* (WSC) for each STT-RAM line to record the write intensity. When the WSC in a line saturates, the cache controller tries to find the line with minimum WSC in the same set as the migration target. If the difference between the two lines is larger than a threshold, migration will be performed.

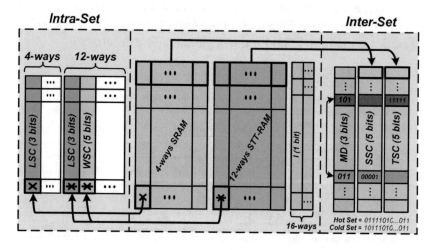

Figure 5.24: Pure dynamic migration scheme: intra-set and inter-set migrations. Taken from [76].

The authors further proposed the inter-set migration to solve the write non-uniformity between cache sets. Eight cache sets that differ in the three most significant bits of their tags form a merge group. The inter-set migration can be performed inside the same merge group. The *STT-RAM saturation counter* (TSC) is used to measure the write intensity of the STT-RAM lines in this set while the *SRAM saturation counter* (SSC) is used for the SRAM lines in this set. The *merge destination* (MD) can be used to build a bidirectional link between the target set (hot set) and victim set (cold set) pair. A STT-RAM line can be migrated to a SRAM line in the same group by using TSC, SSC, and MD. The details of the data replacement algorithm and cache line search algorithm can be found in [76]. Compared to the baseline configuration, the work in [76] can achieve a 49 times improvement in endurance and more than a 50% energy reduction for PARSEC benchmarks.

SW/HW Combined Scheme

The authors in [17] utilize both architectural support and compiler hints to reduce energy and improve endurance of a hybrid cache. The combined scheme improves hybrid cache endurance and energy through both (1) the initial cache block placement and (2) dynamic migration recovery for incorrect initial placement.

The combined scheme leverages the benefits from both compiler analysis and hardware-based runtime support. Static compiler analysis can analyze the memory access patterns in advance and thus has a global view of optimizing block placement. However, there are three limitations to static analysis. First, a compiler cannot accurately capture dynamic behaviors, such as the physical address mapping and cache block replacement. Second, the instrumentation for providing compiler hints has its limitations. It is difficult to provide hints for precompiled libraries or system calls at the kernel level. Third, it is difficult to provide an accurate analysis if the codes are not quite regular. Therefore, Chen et al. utilized hardware support to capture dynamic behaviors [17].

For initial block placement, the decision is made based on both the compiler hints and also the SRAM/STT-RAM capacity pressure monitored in the hardware. To assess the SRAM/STT-RAM capacity pressure, the authors in [17] introduced two additional hardware structures: missing tags (MTs) and MT counters. The proposed structures are similar to the missing tags used in the AH-cache [32] and victim tags used in [16, 137]. The initial block placement policy is described as follows. The compiler hints are obtained from the IR-level codes through LLVM IR analysis and instrumentation. The read and write behaviors of all variables, especially for the array variables, are analyzed at the IR level. A two-bit hint is then embedded in the instruction and is sent to the cache controller. The two-bit hint includes three different compiler predictions: "frequent writes," "infrequent writes," and "unknown." Given both the capacity pressure on two data arrays and the compiler hints from the load/store instructions, the cache controller follows the policy illustrated in Table 5.2 for initial block placement. In general, when the SRAM has high capacity pressure, the strategy is to avoid extra cache misses and is thus performance-driven. When the SRAM has low capacity pressure, the strategy is to avoid the writes on STT-RAM blocks to improve endurance and reduce write energy.

Even with the hardware support, the initial placement may still be incorrect. Therefore, dynamic migration is used to recover the incorrect initial placement, which can avoid excessive writes on a specific block. The migration scheme is similar to the intra-set migration between SRAM and STT-RAM blocks, which was discussed in Section 5.5.2 (in the work [76]). According to [17], results showed that the combined scheme improves the endurance by 23.9x and

Table 5.2: Initial placement policy based on compiler hints and SRAM/STT-RAM capacity pressure. Taken from [17].

Capacity pressure		Compiler hint		
SRAM	STT-RAM	infrequent writes	frequent writes	unknown
High	High	STT-RAM	SRAM	SRAM
High	Low	STT-RAM	STT-RAM	STT-RAM
Low	High	SRAM	SRAM	SRAM
Low	Low	SRAM	SRAM	SRAM

5.9x compared to pure static and pure dynamic schemes, respectively. Furthermore, the system energy can be reduced by 17% compared to a pure dynamic scheme.

Dynamic Placement and Migration Scheme

In [131] the authors proposed a low-cost adaptive placement and migration policy to map write-intensive blocks to SRAM lines in a hybrid cache. The technique considers both initial placement and migration together, as described in [17]. The authors categorized the LLC write accesses into three distinct classes: *core-write*, *prefetch-write*, and *demand-write*. Core-write is a write from the core. For a write-through core cache, it is the write directly from the core to the LLC. For a write-back core cache, it is dirty data evicted from the core cache and written back to the LLC. Prefetch-write is a write from the LLC replacement caused by a prefetch miss. Demand-write is a write from the LLC replacement caused by a demand miss.

In addition to the three write classes, the authors further proposed the *read-range* and *depth-range* to characterize the memory access patterns. The read-range (RR) is used for demand-write and prefetch-write. It is the largest interval between consecutive reads of the block from the time it is fetched into LLC to the time it is evicted. The depth-range (DR) is used for core-write. It records the largest interval between accesses to the block from the current core-write access to the next core-write access. Figure 5.25 shows an example of the behavior of a cache block *a* entering an eight-way set-associative cache until it is evicted. With the definition of read-range and depth-range, the authors further classify the read/depth-range into three types: (1) zero-read/depth-range, (2) immediate-read/depth-range, and (3) distant-read/depth-range. For zero-read/depth range, a block is never read/written again before the eviction. If the read/depth-range is smaller than or equal to two, it is called immediate. Otherwise it is called distant.

Figure 5.26 shows the distribution of access patterns for each type of LLC access. The dynamic placement and migration policy is designed based on Figure 5.26. For the prefetch-write accesses, zero-read-range and immediate-read-range prefetch-write blocks account for 82.5% of all prefetch-write blocks. Therefore, the prefetch-write blocks should be initially placed in SRAM. Once a block is evicted from the SRAM, if it is a distant-read-range block, i.e., it is still live, it

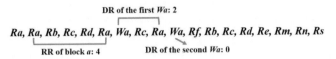

Figure 5.25: An example for illustrating read-range and depth-range. Taken from [131].

should be migrated to STT-RAM lines. Otherwise, the block is dead and should be evicted from the LLC.

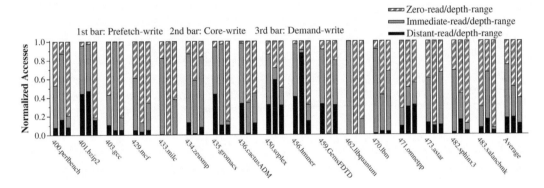

Figure 5.26: The distribution of access pattern for each type of LLC write access. Taken from [131].

The zero-depth-range core-write accesses should remain in this original cache line for avoiding read misses and block migrations. The immediate-depth-range core-write accesses are the write-intensive accesses with write burst behavior. Therefore, it is preferable that they are placed in the SRAM line. The distant-depth-range core-write accesses should remain in the original cache line for minimizing the migration overhead as well.

The zero-read-range blocks account for 61.2% of the demand-write blocks. They are known in the literature as "dead-on-arrival" blocks, and are never referenced again before being evicted. It is unnecessary to place the zero-read-range blocks into LLC, so the blocks should bypass the cache (assuming a non-inclusive cache). For immediate-read-range and distant-read-range demand-write blocks, the authors suggest placing them in the STT-RAM ways to utilize the large capacity of the STT-RAM portion and to reduce the pressure on the SRAM portion. The reader can refer to [131] for the required architectural support and the policy flow chart for dynamic placement and migration. According to [131], the hybird LLC can outperform the SRAM-based LLC on average by 8.0% for single-thread workloads and 20.5% for multicore workloads. The technique reduces power consumption in the LLC by 18.9% and 19.3% for single-thread and multicore workloads, respectively.

CHAPTER 6

Interconnect Customization

6.1 INTRODUCTION

After we discuss the customization of computing units and memory systems, we continue with the communication infrastructure between them. This is a key component since the interconnect latency and bandwidth directly determine whether the computing units and memory system can achieve their designed peak performance. A suboptimal interconnect design can make the customized computing units and the memory system underutilized, and can lead to the corresponding waste of chip area and energy consumption. Fortunately the interconnects have a high potential for improvement from customization as well. We observe that the major difference between a variety of applications usually lies in their data access patterns, while the computing functions needed by these application are quite similar (such as addition and multiplication). Different data access patterns lead to different optimizations on the interconnect topologies and routing policies.

An interconnect infrastructure can be customized from three aspects. First, the interconnect topology can be optimized during the chip design time based on the analysis over the target applications. Second, the routing policy can be optimized during execution based on the compilation and runtime information of the running applications. Third, driven by emerging technologies, the underlying interconnect medium can be optimized by matching the physical property with the application properties. The following sections of this chapter offer a detailed discussion.

6.2 TOPOLOGY CUSTOMIZATION

Depending on the target scenario, the design methodologies of topology customization have the following categories. The first category is application-specific topology synthesis [99, 106]. It is used in the design of application-specific circuits where an irregular interconnect infrastructure is used and fully customized to only one application. The second category is reconfigurable shortcut insertion [10, 93, 100, 126]. This is used in the design of customized domain-specific processors where the interconnects are still based on a regular topology, but are augmented with additional shortcuts at certain positions to adapt to the communication pattern of an application within the target domain. The third category is partial crossbar synthesis and reconfiguration. It is used in the design of accelerator-rich processors. During the design time, we customize the interconnect topology based on the different behaviors of accelerators. Then during runtime, we reconfigure the interconnect topology based on what set of accelerators are turned on by user applications.

6.2.1 APPLICATION-SPECIFIC TOPOLOGY SYNTHESIS

The input of application-specific topology synthesis is usually the communication constraint graph or the application characterization graph. In these kinds of graphs, each vertex represents a core, and each directed edge characterizes the data transfer from one vertex to another vertex. The communication volume and the required bandwidth between each pair of vertices are denoted by the edge weights. The output of the synthesis flow is a network topology which is customized to this communication graph of the target application. The goal is to maximize the throughput or minimize the energy consumption of the target application. Two examples of solving this problem are presented in [106] and [99]. In [106] the authors propose a novel approach to designing the interconnect for a system of computing units, whose interaction is specified from an abstract point of view, as a collection of communication requirements on a set of point-to-point unidirectional "virtual" channels. By abstracting away the specific functionality of each module, they can focus on exploring the various communication topologies that can be built which comprise a set of library elements that include "passive elements" (links) as well as active ones (repeaters, switches), each of them comes with a fixed cost function that captures an application-specific optimality criterion. Then they present two efficient heuristics for the topology synthesis that is based on a decomposition of the optimization problem into two steps: (1) they use quadratic programming to compute quickly the cost of satisfying a set of arc constraints with a shared communication medium, and (2) they propose two clustering algorithms to single out sets of constraints that should be considered together. In [99] the authors analyze the frequently encountered generic communication primitives, such as gossiping (all-to-all communication), broadcasting (one-to-all), and multicasting (one-to-many). They propose an approach based on using these primitives as an alphabet capable of characterizing any given communication pattern. The proposed algorithm searches through the entire design space for a solution that minimizes the system total energy consumption, while satisfying the other design constraints. Compared to the regular mesh architecture, the customized interconnect infrastructure generated by their approach shows about 36% throughput increase and 51% reduction for the benchmark Advanced Encryption Standard (AES).

Although in a CSoC platform, it may not be possible to synthesize and deploy an application-specific topology for each application, one may embed an application-specific topology in a generic (say mesh-like) NoC for application-specific optimization. One example is the work on express virtual channels (EVC) [83], which presented an effective flow control mechanism which allows packets to virtually bypass intermediate routers along their path in a non-speculative fashion, resulting reduced the energy/delay toward that of a dedicated wire while simultaneously improving the NoC throughput. In a recent work [30] the authors made the observation that in accelerator-rich architectures, accelerator-memory accesses exhibit predictable patterns, creating highly utilized network paths for each application. Therefore, they propose a technique of reserving NoC paths based on the timing information from the global accelerator manager, with additional regulation of the communication traffic through TLB buffering and hybrid-switching. The combined effect of these optimizations lead to total execution time reduction over a both the baseline packet-switched mesh NoC and the improved version of NoC using the EVC optimization technique.

6.2.2   RECONFIGURABLE SHORTCUT INSERTION

Adaptive shortcuts allow us to selectively provide bandwidth to an application's critical communications, enabling us to retain a high level of performance with a much simpler underlying conventional wire mesh. Shortcut modules can be integrated on top of a network-on-chip, providing single-cycle shortcuts that accelerate communication from a set of source routers to a set of destination routers. We refer to this set of source and destination routers as shortcut-enabled routers. Note that these kinds of routers can be implemented more efficiently by advanced physical media, including radio-frequency devices and optical devices; these will be discussed in Section 6.4. In a mesh topology for example, standard routers have five input/output ports, which carry traffic to/from their north, south, east, and west neighbors, as well as to a local computing element like a cache or core (attached to the fifth port). To add shortcuts into a mesh, each shortcut-enabled router must be given a sixth port, which connects it to either transmitter (if it is a source router), receiver (if it is a destination router), or both (if it is both sending and receiving on shortcuts). When the mesh is extended to include shortcuts, one needs to switch from the XY routing to the shortest-path routing. To realize a reconfigurable network-on-chip, the set of shortcuts present in the network must be changed such that the set of source and destination routers is modified to match the current communication demand of network traffic. The basic way to achieve this is to tune the selected transmitter and receiver at each shortcut-enabled router to send and listen on the same frequency band to establish a shortcut connection. The flexibility of reconfiguration does come with two costs: arbitration for frequency bands among transmitters and receivers, and subsequent integration of the resulting shortcuts into network packet routing.

The work in [10] assumes a coarse-grain approach to arbitration, where shortcuts are established for the entire duration of an application's execution. This allows one to amortize the cost of reconfiguration over a large number of cycles. A reconfiguration of the architecture involves the following steps: 1) Shortcut selection—one must decide which shortcuts will augment the underlying topology. This can be done ahead of time by the application writer or compiler, or at run time by the operating system, a hypervisor, or in the hardware itself. 2) Transmitter/receiver tuning—based on shortcut selection, each transmitter or receiver in the topology will be tuned to a particular configuration (or disabled entirely) to implement the shortcuts. 3) Routing table updates—new routes must be established and programmed into the routing tables at all network routers to match the new available paths. If all network routers are updated in parallel, and each routing table has a single write port, it would take 99 cycles to update all the routes in the network (1 cycle for each other router in the network). Since this work considers per-application reconfiguration of an NoC, this cost can easily be overlapped with other context switch activity, and will not increase the start delay when executing an application.

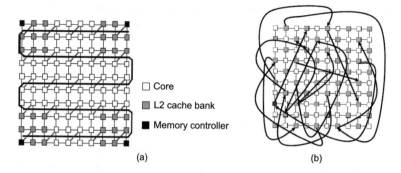

Figure 6.1:   (a) Shortcut routing channels overlaid with conventional mesh interconnects, (b) adaptive shortcuts for 1Hotspot trace. Taken from [10].

Figure 6.1(a) demonstrates a conventional mesh topology with a set of overlaid shortcut routing channels. Here, this work constrains the number of shortcut-enabled routers to half of the total routers (50 routers). In this figure the routers appear to have a small diagonal connection to the set of shortcut routing channels, which is represented as a single thick line winding through the mesh.

To reconfigure the set of shortcuts dynamically for each application (or per workload), application communication statistics can be introduced into the cost equation. Intuitively, the goal is to accelerate communication on paths that are most frequently used by the application, operating under the assumption that these paths are most critical to application performance. To identify these paths, this work needs to rely on information that can be readily collected by event counters in the network. The metric it uses to guide the selection is inter-router communication frequency. From a given router X to another router Y, communication frequency is measured as the number of messages sent from X to Y. To determine the maximum benefit of this approach, it assumes that this profile is available for the applications it wishes to run. Then the target of the shortcut selection algorithm is to minimize $\sum_{x,y} F_{x,y} W x, y$ where $F_{x,y}$ is the total number of messages sent from router x to router y, and $W_{x,y}$ is the length of the shortest path between x and y. This can be solved using the heuristic approach in [10]. An example of application-specific shortcut selection is shown for the 1Hotspot trace in Figure 6.1(b). Experimental results in [10] show that adaptive shortcut insertion can enable a 65% NoC power savings while maintaining comparable performance.

6.2.3   PARTIAL CROSSBAR SYNTHESIS AND RECONFIGURATION

The discussion above mainly targets enhancing conventional interconnects designed for CPU-centric architectures. The effectiveness of these interconnects is based on the assumption that each CPU core performs a load/store every few clock cycles. Therefore, a simple interconnect, e.g., Fig. 6.2(a), can arbitrate the data channel alternatively among CPU cores without reducing much of their performance.

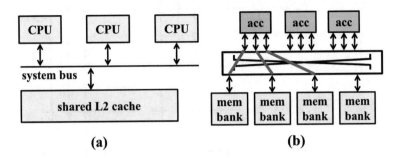

Figure 6.2:   Difference between memory sharing among general-purpose CPU cores and among accelerators. (a) A simple interconnect for CPU cores, (b) demanding interconnects for accelerators.

In contrast, accelerators may run >100x faster than CPUs [67], and each accelerator needs to perform several loads/stores every clock cycle. The interconnects between accelerators and shared memories need to be high-speed, high-bandwidth and contain many conflict-free data channels to prevent accelerators from being starved for data, as shown in Fig. 6.2(b). An accelerator needs to have at least n ports if it wants to fetch n data every cycle. The n ports of the accelerator need to be connected to n memory banks via n conflict-free data paths in the interconnects.

Another problem with conventional interconnect designs is that the interconnect arbitration among requesting accelerators is performed upon each data access. NoC designs even perform multiple arbitrations in a single data access as the data packet goes through multiple routers. Since accelerators implement computation efficiently, but have no way of reducing the number of necessary data accesses, the extra energy consumed by the interconnect arbitration during each data access will become a major concern. The arbitration upon each data access also leads to a large and unexpected latency for each access. Since accelerators aggressively schedule many operations (computation and data accesses) into every time slot [36], any late response of data access will stall many operations and lead to significant performance loss. Many accelerator designs prefer small and fixed latencies to keep their scheduled performance, and because of this many accelerators [89] have to give up memory sharing.

The work in [40] designs the interconnects into a configurable crossbar as shown in Fig. 6.3.

The configuration of the crossbar is performed only upon accelerator launch. Each memory bank will be configured to connect to only one accelerator (only one of the switches that connect the memory bank will be turned on). When accelerators are working, they can access their connected memory banks just like private memories, and no more arbitration is performed on their data paths. Fig. 6.3 shows that acc 1, which contains three data ports (i.e., demands three data accesses every cycle), is configured to connect memory banks 1–3. Note that there can be other

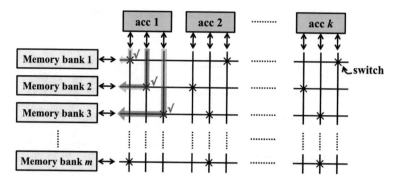

Figure 6.3: Interconnects designed as a configurable crossbar between accelerators and shared memories to keep data access cost small. Taken from [40]. ✓ : switch turned on.

design choices for the configurable network between accelerators and shared memories. However the crossbar design contains the fewest logics (only one switch) in the path from an accelerator port to a memory bank, and thus helps minimize the data access latency. It can achieve an access latency of only one clock cycle in an FPGA prototyping so that the timing of access to shared memories acts exactly as private memories within accelerators.

The primary goal of the crossbar design is that for any set of accelerators that are powered on in an accelerator-rich platform and require t memory banks in total, a feasible configuration of the crossbar can be found to route the t data ports of the accelerators to t memory banks. A full crossbar that connects every accelerator port to every memory bank provides a trivial solution. However, it is extremely area-consuming. One would like to find a sparsely populated crossbar (e.g., Fig. 6.3) to achieve high routability. The definition of routability is given as follows:

Definition 6.1   Suppose the total number of accelerators in an accelerator-rich platform is k; the number of data ports of the accelerators is n; the number of memory banks in the platform is m; and the maximum number of accelerators that can be powered on under power budget in dark silicon is c. The routability of the crossbar is defined as: the probability that a randomly selected workload of c accelerators out of the total k accelerators can be routed to $c \times n$ memory banks via $c \times n$ separate data paths in the crossbar. The goal here is to *optimize the crossbar for the fewest switches while keeping high routability*.

Designers can identify three optimization opportunities that emerge in accelerator-rich platforms, and exploit them to develop novel interconnects between accelerators and shared memories:

- An accelerator contains multiple data ports, and in the interconnect design the relations of the ports from the same accelerator should be handled differently compared to the ports from different accelerators. A two-step optimization can be used rather than optimizing all the ports of all accelerators globally in a single procedure. Many unnecessary connections associated with each individual accelerator are identified and removed before all the accelerators are optimized together.

- Due to the power budget in dark silicon, only a limited number of accelerators will be powered on in an accelerator-rich platform. The interconnects can be partially populated to just fit the data access demand limited by the power budget.

- Accelerators are heterogeneous. Some accelerators will have a higher chance of being turned on or off together if they belong to the same application domain. This kind of information can be used to customize the interconnect design to remove potential data path conflicts and use fewer transistors to achieve the same efficiency.

Experimental results in [40] show that the crossbar customization for accelerators can achieve 15x area savings and 10x performance improvement compared to conventional NoCs that were optimized for CPU cores.

6.3   ROUTING CUSTOMIZATION

Depending on the target scenario, the design methodologies of routing customization have the following categories. The first is application-aware deadlock-free routing [37, 103]. It is used in the design of CPU-centric architectures where data packets transmitted over the interconnects are nondeterministic at the level of clock cycle and we can only optimize the policies of each router when it receives a packet. The second is data flow synthesis [33]. It is used in the design of application-specific accelerators where the data flows can be optimized during design time for the minimum area and energy consumption.

6.3.1   APPLICATION-AWARE DEADLOCK-FREE ROUTING

The evolution of application-specific interconnects leads to irregular topologies. However, efficient deadlock-free routing for interconnects with irregular topologies remains an open problem.

There are two main approaches to dealing with deadlock in irregular interconnects. The first class of approaches is based on the theory of [48]. It divides the NoC into two virtual networks (VN): one fully-adaptive (no routing restrictions) and another that is deadlock-free (with routing restrictions). Network packets are routed in the full-adaptive VN at first and will be moved to the deadlock-free VN when there are no available resources in the full-adaptive VN. One recent example is the deadlock detection and recovery method used in [10]. However, since most CSoCs in the embedded system domain are power-critical, the power overhead of introducing two virtual channels for each physical channel is significant. In addition, this method is applicable to general-purpose computing architectures and wastes the application information.

The second class of approaches handles the deadlock-free routing problem in irregular interconnects by imposing routing restriction, such as the turn prohibition algorithm [120] and south-last routing [100]. They restrict routings without considering the application-specific communication patterns; hence they may increase the routing distance between heavily communicated nodes. The recent work in [103] first proposes to remove dependencies based on the application communication requirement. It uses a greedy heuristic and may exhaustively enumerate all possible combinations of channel dependency cycles and thus cannot scale to large designs. In fact, sometimes it is impossible to make the channel dependency graph acyclic without disconnecting the network with unidirectional links. This complication is not considered in [103]. Moreover, [103] only considers whether there is communication between two nodes or not, without consideration of the data size transferred, which may lead to suboptimal solutions. Therefore, it is necessary to find an optimal trade-off point between power and performance (i.e., between these two types of approaches) for deadlock-free routing in irregular interconnects. We want something that avoids restricting critical routes in the interconnects, but that also does not significantly increase interconnect power. The work in [37] proposed an application-specific cycle elimination and splitting (ACES) method for this problem. It first develops a scalable algorithm using global optimization to eliminate as many channel dependency cycles as possible with the guarantee of network reachability, based on the application-specific communication patterns, and then only splits the remaining small set of cycles (if any) using virtual channel splitting. Network performance is maintained by ensuring plentiful shortest paths between heavily communicated nodes. Moreover, with the possible existence of split channels, a routing table construction and encoding method is developed to minimize the hardware overhead of ACES.

An overview of the ACES framework is shown in Figure 6.4(a). The framework takes the application characterization graph (APCG) and topology graph (TG) as the inputs, as shown in Figure 6.4(b). A channel dependency graph (CDG) is constructed based on the given TG without any routing restrictions, as shown in Figure 6.4(c). From the APCG and CDG, an initial application-specific channel dependency graph (ASCDG) is constructed by weighting the channel dependencies with the application-specific communication patterns, as shown in Figure 6.4(d). Then a heuristic algorithm is performed to remove channel dependencies that are not part of frequent communication routes. If the algorithm finishes with an acyclic ASCDG, virtual channel splitting is bypassed. Otherwise, it must be the case that the channel dependencies in the remaining cycles are kept to maintain the network reachability; thus, virtual channel splitting is used to break these remaining cycles. Furthermore, since application-specific interconnects typically use routing tables to guide the routers to route the network packets [57], one can construct and encode the routing tables based on the final acyclic ASCDG to minimize the hardware overhead of ACES. At the run time, the generated routing table will be loaded into the routers at the beginning of the application. The comparison between Figure 6.4(e) and Figure 6.4(f) shows the benefit of ACES. Without consideration of the application communication pattern, an acyclic CDG generated by the south-last routing algorithm is given in Figure 6.4(e), where the use of the shortcut is restricted with the forbidden southbound dependency from AL to LH/LF. The shortcut from A to L is augmented to optimize the topology for the application, but by applying

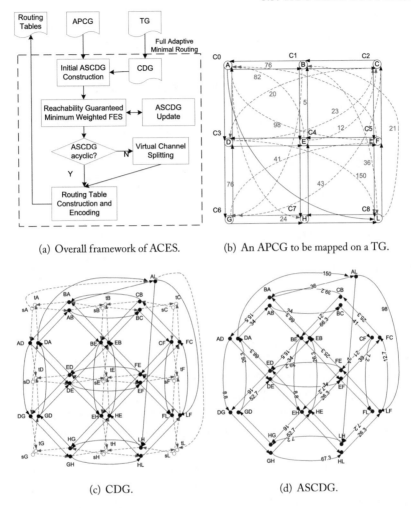

(a) Overall framework of ACES.

(b) An APCG to be mapped on a TG.

(c) CDG.

(d) ASCDG.

Figure 6.4: Examples to illustrate ACES routing optimization. Taken from [37]. *(Continues.)*

south-last routing, the use of the shortcut is severely restricted. Figure 6.4(f) shows the acyclic CDG generated by ACES, where only the channel dependency edges that are never or rarely used are removed. It should be noted that reachability should be guaranteed while breaking cycles in the ASCDG; i.e., for each pair of communicating nodes of the application, there is at least one directed path from the source node to the destination node.

Experimental results in [37] show that ACES can either reduce the NoC power by 11% ∼ 35% while maintaining approximately the same network performance, or improve the network performance by 10% ∼ 36% with slight NoC power overhead (−5% ∼ 7%) on a wide range of examples.

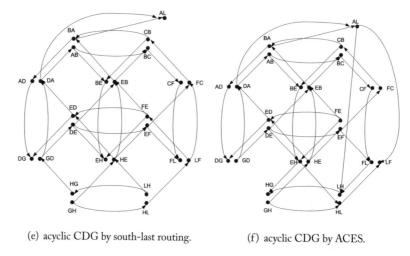

(e) acyclic CDG by south-last routing.   (f) acyclic CDG by ACES.

Figure 6.4: *(Continued.)* Examples to illustrate ACES routing optimization. Taken from [37].

6.3.2 DATA FLOW SYNTHESIS

In application-specific accelerators, data flows on the interconnects can be fully customized to each application. During the design time, one can decide when to fetch data from off-chip, when to move data from one buffer to another, and when to send data to computation units. The synthesis goal is to minimize the off-chip accesses and the on-chip buffering logics without flow conflicts on the interconnects in any clock. The methodology of data flow synthesis varies among different application domains with different data access patterns. We shall illustrate that using a popular application domain—stencil computation.

Stencil computation comprises an important class of kernels in many application domains, such as image processing, constituent kernels in multigrid methods, and partial differential equation solvers. These kernels often contribute to most workloads in these applications. Even in recent technologies on memory partitioning [87, 130] which were developed for general applications, all the benchmarks used are in fact stencil computation.

The data elements accessed in stencil computation are on a large multidimensional grid which usually exceeds on-chip memory capacity. The computation is iterated as a stencil window slides over the grid. In each iteration, the computation kernel accesses all the data points in the stencil window to calculate an output. Both the grid shape and the stencil window can be arbitrary, as specified by the given stencil applications. A precise definition of stencil computation can be found in [34, 70].

```
void denoise2D( float A[768][1024],
                float B[768][1024] )
{
```

```
for ( int i = 1; i < 767; i++ )
    for ( int j = 1; j < 1023; j++ )
        B[ i ][ j ] =
            pow (A[ i ][ j ] − A[ i ][ j −1], 2) +
            pow (A[ i ][ j ] − A[ i ][ j +1], 2) +
            pow (A[ i ][ j ] − A[ i −1][ j ], 2) +
            pow (A[ i ][ j ] − A[ i +1][ j ],  2);
}
```

Listing 6.1: Example C code of a typical stencil computation (5-point stencil window in the kernel 'DENOISE' in medical imaging [38]).

Listing 6.1 shows an example stencil computation in the kernel 'DENOISE' in medical imaging [38].

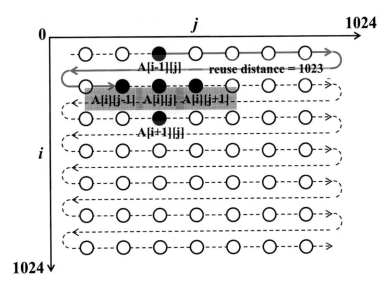

Figure 6.5: Iteration domain of the example stencil computation in Listing 6.1. Taken from [33].

Its grid shape is a 768×1024 rectangle, and its stencil window contains five points, as shown in Fig. 6.5. Five data elements need to be accessed in each iteration. In addition, many data elements will be repeatedly accessed among these iterations. For example, $A[2][2]$ will be accessed five times, when $(i, j) \in \{(1, 2), (2, 1), (2, 2), (2, 3), (3, 2)\}$. This leads to high on-chip memory port contention and off-chip traffic, especially when the stencil window is large (e.g., after loop fusion of stencil applications for computation reduction as proposed in [97]). Therefore, during the hardware development of a stencil application, a large portion of engineering effort is spent on data reuse and memory partitioning optimization.

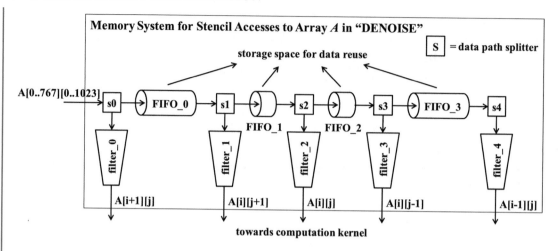

Figure 6.6: The example circuit structure of the memory system generated for array A in the stencil computation of Listing 6.1. Taken from [33].

The work in [33] uses a chain structure of the interconnects, as illustrated in the example in Fig. 6.6, which is generated for the stencil computation in Listing 6.1. Suppose the stencil window contains n points ($n = 5$ in the example of Listing 6.1). The memory system will contain $n - 1$ data reuse FIFOs as well as n data path splitters and n data filters connected together in the way shown in Fig. 6.6. The data reuse FIFOs provide the same storage as conventional data reuse buffers, and the data path splitters and filters work as memory controllers and data interconnects. The $n - 1$ buffers and the n routers are the theoretical lower-bound of the module count to satisfy n data accesses in the stencil window every clock cycle. After the data flow synthesis, the data required by each data access port of the computation kernel can be provided by each FIFO which receives data from its precedent FIFO at the same time. Interconnect contention is completely eliminated here. Experimental results show $25 - -66\%$ area savings while maintaining the same network performance.

6.4   CUSTOMIZATION ENABLED BY NEW DEVICE/CIRCUIT TECHNOLOGIES

Depending on the underlying physical materials, the medium customization have the following three categories. The first is based on the emerging nanophotonics technology. The second is based on radio-frequency interconnects. The third is based on non-volatile memory switches.

6.4.1   OPTICAL INTERCONNECTS

Nanophotonics offers an opportunity to reduce the power and area of off- and on-stack interconnects while meeting future system bandwidth demands. Optics is ideal for global communication because the energy cost is incurred only at the endpoints and is largely independent of length. Dense wavelength division multiplexing (DWDM) enables multiple single-wavelength communication channels to share a waveguide, providing a significant increase in bandwidth density. Recent nanophotonics developments demonstrate that waveguides and modulation/demodulation circuit dimensions are approaching electrical buffer and wire circuit dimensions [88].

The authors of [127] presented Corona, a 3D many-core NUMA system that uses nanophotonic communication for both inter-core communication and off-stack communication to memory or I/O devices. A photonic crossbar fully interconnects its 256 low-power multi-threaded cores at 20 terabyte per-second bandwidth. The crossbar enables a cache coherent design with near uniform on-stack and memory communication latencies. Photonic connections to off-stack memory enable unprecedented bandwidth to large amounts of memory with only modest power requirements.

6.4.2   RADIO-FREQUENCY INTERCONNECTS

Radio frequency interconnect (or RF-I) was proposed as a high aggregate bandwidth, low latency alternative to traditional interconnects. Its benefits have been demonstrated for off-chip, on-board communication as well as for on-chip interconnection networks [11]. On-chip RF-I is realized via transmission of electromagnetic waves over a set of transmission lines, rather than the transmission of voltage signals over a wire. When using traditional voltage signaling, the entire length of the wire has to be charged and discharged to signify either '1' or '0', consuming much time and energy. In RF-I an electromagnetic carrier wave is continuously sent along the transmission line instead. Data is modulated onto that carrier wave using amplitude and/or phase changes. It is possible to improve bandwidth efficiency of RF-I by sending many simultaneous streams of data over a single transmission line. This is referred to as multiband (or multicarrier) RF-I. In multiband RF-I, there are N mixers on the transmitting (or Tx) side, where N is the number of senders sharing the transmission line. Each mixer up-converts individual data streams into a specific channel (or frequency band). On the receiver (Rx) side, N additional mixers are employed to down-convert each signal back to the original data, and N low-pass-filters (LPF) are used to isolate the data from residual high-frequency components. RF-I has been projected to scale better than traditional RC wires in terms of delay and power consumption, and unlike traditional wires, it can allow signal transmission across a $400mm^2$ die in 0.3 ns via propagation at the effective speed of light, as opposed to less than or equal to 4 ns on a repeated bus. Chang et al. [11] used RF-I transmission lines on a 64-core CMP to realize shortcuts on a mesh interconnect. They explored the potential of adaptive-routing techniques to avoid bottlenecks resulting from contention for the shortcuts.

6.4.3   RRAM-BASED INTERCONNECTS

In customizable architectures, computing units, memory systems and interconnects will be reconfigured to compose the accelerator for the target application to be executed. After reconfiguration, the data paths over the interconnects are fully customized and keep constant during the whole execution of the application. The routers in this kind of interconnect act as programmable switches which can be configured to certain routing directions instead of dynamic routing upon each data packet. In conventional CMOS technology, SRAM-based pass transistors are a good candidate of the router implementation.

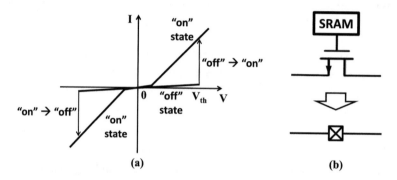

Figure 6.7:   Illustration of NVM-based switches taken from [41]. (a) Hysteresis characteristic of a two-terminal NVM device, (b) function as a routing switch in place of a pass transistor and its six-transistor SRAM cell [13, 59, 60, 101, 122].

Driven by emerging non-volatile memories (NVMs), they can also be replaced by NVM-based switches, as shown in Figure 6.7(b). This kind of use is enabled by a common property of these emerging NVMs. That is, the connection between two terminals of these devices can be programmed to turn on or turn off, as shown in Figure 6.7(a). By applying specific programming voltages, the resistance between the two terminals can be switched between the "on" state and the "off" state. The programmed resistance value can be kept either under operating voltages or without supply voltage due to nonvolatility. This kind of NVM use saves the area of SRAMs and also the pass transistors that build routing switches. An example provided in [41] uses an interconnect architecture based on resistive RAMs (RRAMs).

In [41], the programmable interconnects are composed of three disjoint structures:

- Transistor-less programmable interconnects

- A programming grid

- An on-demand buffering architecture

This composition takes routing buffers from programmable interconnects and puts them in a separate architecture. The transistor-less programmable interconnects correspond to SRAM-based

configuration bits and MUX-based routing switches in conventional programmable interconnects. They are built by RRAMs and metal wires alone and are placed over CMOS transistors, as
shown in Fig. 6.8.

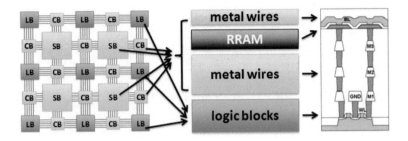

Figure 6.8: Switch blocks and connection blocks in transistor-less programmable interconnects are
placed over logic blocks in the same die according to existing RRAM fabrication structures [65, 117,
124, 129]. Taken from [41].

In the structure of the transistor-less programmable interconnects, RRAMs and metal wires
are stacked over CMOS transistors. The layout will be very different from that of conventional
programmable interconnects and will be applied with very tight space constraints. This work provides an RRAM-friendly layout design which solves these constraints and at the same time fits
into the footprint of the CMOS transistors below. The programming transistors in the programming grid are heavily shared among RRAMs via the transistor-less programmable interconnects.
The on-demand buffering architecture provides opportunities to allocate buffers in interconnects
during the implementation phase. It allows utilization of the application information for a better allocation of buffers. Note that the feasibility of the disjoint structures, and the feasibility
of all their improvements mentioned above, is based on the use of RRAMs as programmable
switches. Simulation results show that the RRAM-based programmable interconnects achieve a
96% smaller footprint, 55% higher performance, and 79% lower power consumptions.

CHAPTER 7

Concluding Remarks

Customizable SoC processors have tremendous promise for meeting the future scaling requirements of application writers while adapting to the scaling challenges anticipated in future technology nodes. Customization can benefit all levels of CSoC design, including cores, memory, and interconnect.

Customized cores and compute engines can dramatically reduce power dissipation by reducing or even eliminating unnecessary non-computational sub-components. Preliminary work in this area has demonstrated cores that can be customized with specialized instructions or components. Further work has shown that accelerator-rich designs can outperform general-purpose processor designs, but can still retain some adaptability and flexibility to enhance design longevity through either compile-time reconfiguration or runtime composition. Future work in this area may explore migrating computation closer to memory, building customized cores and accelerators in or near main memory or disk. Such embedded accelerators can help to not only reduce communication latency and power, but can customize components external to the SoC.

Customized on-chip memory includes the specialization of memory resources for an application's memory requirements from reducing energy hungry associative ways to careful software-directed placement of key memory blocks to reduce cache misses. Preliminary work has begun to expand upon the conventional memory resources available to the CSoC with emerging memory technology that features various power and performance trade-offs. Future work may provide a set of different memory resources that can be flexibly mapped to an application depending on resource requirements. For example, memory with low leakage power can be used for higher capacity read-only memory.

Interconnect customization allows adaptation of the on-chip network to the communication patterns of a particular application. While preliminary work has demonstrated gains in topology customization and routing customization, there is still tremendous potential in alternative interconnects that can provide high-bandwidth, extremely low latency communication across large CSoCs. Such interconnect has potential to accelerate critical communication primitives in parallel code, such as locks, mutexes, and barriers. With faster synchronization and communication, we may find that more applications can efficiently leverage the growing number of cores and accelerators on CSoCs. Moreover, with growing computational power from customized cores, alternative interconnects may hold the key to supplying sufficient bandwidth to feed these high performing cores.

Most of these techniques were evaluated using simulators. To further validate the CSoC concept, recent work in [15, 18] prototypes a real ARA on the Xilinx Zynq-7000 SoC [133] with four medical imaging accelerators. The Zynq SoC, is composed of a dual-core ARM Cortex-A9 and FPGA fabrics, which can be used to realize the accelerators, interconnects and on-chip shared memories in a CSoC. Table 7.1 shows the performance and power results of *denoise* application in the ARA prototype and the state-of-the-art processors [18]. The prototype can achieve 7.44x and 2.22x energy efficiency over the state-of-the-art processors, Xeon and ARM, respectively. As reported in [85], the power gap between FPGA and ASIC is around 12X. If the ARA is implemented in ASIC, a 24 to 84X energy saving over Xeon processors is expected. This is an encouraging step to further validate the effectiveness of a CSoC.

Table 7.1: Performance and power comparison over (1)ARM Cortex-A9, (2)Intel Xeon (Haswell), and (3)ARA

	Cortex-A9	Xeon (OpenMP, 24 threads)	ARA
Freq.	667MHz	1.9GHz	Acc@100MHz CPU@667MHz
Runtime(s)	28.34	0.55	4.53
Power	1.1W	190W(TDP)	3.1W
Total Energy	2.22x	7.44x	1x

Although the SoC-level integration of customized accelerators and multicore processors is still at an early stage (the Xilinx Zynq-7000 FPGA SoC [133] is a good example in this direction), the leading companies, such as Intel and IBM, have started to integrate multicore processors with FPGA fabrics at the server level. Intel's QuickAssist acceleration technology [73] and IBM's coherent accelerator processor interface (CAPI) for POWER8 [72] are two examples of industry adoption of customized computing. These products both allow users to offload compute-intensive kernels in their applications to FPGA-based accelerators.

QuickAssist is a two-socket system, where an Intel Xeon processor is located in one socket and an Altera FPGA is located in the other. The applications launched on the host Xeon processor can exchange data with an accelerator function unit (AFU) implemented on the FPGA through the Intel QPI. The FPGA data can be kept coherent with the last-level cache (LLC) in the Xeon processor. To achieve this, the FPGA implements a QPI caching agent to communicate with the LLC, which means that it can access all of the main memory connected to the Xeon processor. The corresponding APIs are also provided for users to develop their applications.

IBM CAPI also allows users to offload computation-heavy tasks to specialized accelerators. The major difference between CAPI and QuickAssist is that the CAPI system uses a PCIe-based FPGA coprocessor. The CAPI system also provides coherent cache support. The coherent accelerator processor proxy (CAPP) unit in the POWER8 processor maintains a directory of all cache lines held by the off-chip accelerators. On the FPGA side, a POWER service layer (PSL) is built to communicate with the POWER8 processor. An application can be the master or the slave to the accelerator based on the demand.

In fact, while this lecture focuses on customization of on-chip components, there is also a lot of exciting work using customization and specialization beyond the single-chip level. For example, FPGAs and GPUs are widely used as co-processor accelerators for many applications (e.g., [22, 69, 90]). Recently, the Microsoft Catapult project introduced over 1,600 CPU servers with FPGAs to accelerate the Bing search engine [109] at the datacenter scale. Also, there is various work looking to move accelerators into or close to DRAMs or storage systems for acceleration (e.g., [46, 102]). Finally, customizable interconnects, such as RF-Interconnect, can also be used for off-chip communication (e.g. [79]).

Clearly, customization can be done at multiple levels—chip level, server-node level, or even data center level. There are plenty of research opportunities at each level. For example, at the chip level, it is an open problem how to decide the best set of accelerator building blocks (ABBs) automatically for a given application domain. The ABBs used in Section 4.4.3 were chosen manually. At the server-node level, the benefits and overhead of providing transparent coherent support between the host and accelerator memories are still not very understood. At the data center level, an intelligent runtime resource management system is necessary to enable efficient sharing of accelerators between different jobs or different tasks within the same job. We hope that this Synthesis lecture serves as a starting point to stimulate more new research in this exciting area.

Bibliography

[1] D. H. Albonesi. Selective cache ways: On-demand cache resource allocation. In *Proceedings of the 32Nd Annual ACM/IEEE International Symposium on Microarchitecture*, MICRO 32, pages 248–259, Washington, DC, USA, 1999. IEEE Computer Society. DOI: 10.1109/MICRO.1999.809463. 41, 43, 44, 60, 61

[2] K. Atasu, O. Mencer, W. Luk, C. Ozturan, and G. Dundar. Fast custom instruction identification by convex subgraph enumeration. In *Application-Specific Systems, Architectures and Processors, 2008. ASAP 2008. International Conference on*, pages 1–6. IEEE, 2008. DOI: 10.1109/ASAP.2008.4580145. 20, 23

[3] T. M. Austin and G. S. Sohi. Zero-cycle loads: Microarchitecture support for reducing load latency. In *Proceedings of the 28th Annual International Symposium on Microarchitecture*, MICRO 28, pages 82–92, Los Alamitos, CA, USA, 1995. IEEE Computer Society Press. DOI: 10.1109/MICRO.1995.476815. 52

[4] A. Baniasadi and A. Moshovos. Instruction flow-based front-end throttling for power-aware high-performance processors. In *Proceedings of the 2001 international symposium on Low power electronics and design*, pages 16–21. ACM, 2001. DOI: 10.1145/383082.383088. 16

[5] C. F. Batten. *Simplified vector-thread architectures for flexible and efficient data-parallel accelerators*. PhD thesis, Massachusetts Institute of Technology, 2010. 20

[6] C. J. Beckmann and C. D. Polychronopoulos. Fast barrier synchronization hardware. In *Proceedings of the 1990 ACM/IEEE conference on Supercomputing*, pages 180–189. IEEE Computer Society Press, 1990. DOI: 10.1109/SUPERC.1990.130019. 19

[7] S. Borkar and A. A. Chien. The future of microprocessors. *Communications of the ACM*, 54(5):67–77, May 2011. DOI: 10.1145/1941487.1941507. 1, 2

[8] A. Buyuktosunoglu, T. Karkhanis, D. H. Albonesi, and P. Bose. Energy efficient co-adaptive instruction fetch and issue. In *Computer Architecture, 2003. Proceedings. 30th Annual International Symposium on*, pages 147–156. IEEE, 2003. DOI: 10.1145/871656.859636. 16

[9] C. Bienia et al. The PARSEC benchmark suite: Characterization and architectural implications. Technical Report TR-811-08, Princeton University, 2008. DOI: 10.1145/1454115.1454128. 2

[10] M.-C. F. Chang, J. Cong, A. Kaplan, C. Liu, M. Naik, J. Premkumar, G. Reinman, E. Socher, and S.-W. Tam. Power reduction of CMP communication networks via RF-interconnects. *2008 41st IEEE/ACM International Symposium on Microarchitecture*, pages 376–387, Nov. 2008. DOI: 10.1109/MICRO.2008.4771806. 69, 71, 72, 75

[11] M. F. Chang, J. Cong, A. Kaplan, M. Naik, G. Reinman, E. Socher, and S.-W. Tam. CMP network-on-chip overlaid with multi-band RF-interconnect. *2008 IEEE 14th International Symposium on High Performance Computer Architecture*, pages 191–202, Feb. 2008. DOI: 10.1109/HPCA.2008.4658639. 81

[12] C. Chen, W. S. Lee, R. Parsa, S. Chong, J. Provine, J. Watt, R. T. Howe, H. P. Wong, and S. Mitra. Nano-Electro-Mechanical Relays for FPGA Routing : Experimental Demonstration and a Design Technique. In *Design, Automation and Test in Europe Conference and Exhibition (DATE)*, 2012. DOI: 10.1109/DATE.2012.6176703. 42

[13] C. Chen, H.-S. P. Wong, S. Mitra, R. Parsa, N. Patil, S. Chong, K. Akarvardar, J. Provine, D. Lewis, J. Watt, and R. T. Howe. Efficient FPGAs using Nanoelectromechanical Relays. In *International Symposium on FPGAs*, pages 273–282, 2010. DOI: 10.1145/1723112.1723158. 82

[14] Y. Chen, W.-F. Wong, H. Li, and C.-K. Koh. Processor caches built using multi-level spin-transfer torque ram cells. In *Low Power Electronics and Design (ISLPED) 2011 International Symposium on*, pages 73–78, Aug 2011. DOI: 10.1109/ISLPED.2011.5993610. 58

[15] Y.-T. Chen, J. Cong, M. Ghodrat, M. Huang, C. Liu, B. Xiao, and Y. Zou. Accelerator-rich cmps: From concept to real hardware. In *Computer Design (ICCD), 2013 IEEE 31st International Conference on*, pages 169–176, Oct 2013. DOI: 10.1109/ICCD.2013.6657039. 86

[16] Y.-T. Chen, J. Cong, H. Huang, B. Liu, C. Liu, M. Potkonjak, and G. Reinman. Dynamically reconfigurable hybrid cache: An energy-efficient last-level cache design. In *Proceedings of the Conference on Design, Automation and Test in Europe*, DATE '12, pages 45–50, San Jose, CA, USA, 2012. EDA Consortium. DOI: 10.1109/DATE.2012.6176431. 42, 58, 59, 60, 61, 62, 65

[17] Y.-T. Chen, J. Cong, H. Huang, C. Liu, R. Prabhakar, and G. Reinman. Static and dynamic co-optimizations for blocks mapping in hybrid caches. In *Proceedings of the 2012 ACM/IEEE International Symposium on Low Power Electronics and Design*, ISLPED '12, pages 237–242, New York, NY, USA, 2012. ACM. DOI: 10.1145/2333660.2333717. 42, 58, 59, 60, 61, 63, 65, 66

[18] Y.-T. Chen, J. Cong, and B. Xiao. Aracompiler: a prototyping flow and evaluation framework for accelerator-rich architectures. In *Performance Analysis of Systems and Software (ISPASS), 2015 IEEE International Symposium on*, pages 157–158, March 2015. DOI: 10.1109/ISPASS.2015.7095795. 86

[19] E. Chi, A. M. Salem, R. I. Bahar, and R. Weiss. Combining software and hardware monitoring for improved power and performance tuning. In *Interaction Between Compilers and Computer Architectures, 2003. INTERACT-7 2003. Proceedings. Seventh Workshop on*, pages 57–64. IEEE, 2003. DOI: 10.1109/INTERA.2003.1192356. 16

[20] D. Chiou, P. Jain, L. Rudolph, and S. Devadas. Application-specific memory management for embedded systems using software-controlled caches. In *Proceedings of the 37th Annual Design Automation Conference*, DAC '00, pages 416–419, New York, NY, USA, 2000. ACM. DOI: 10.1145/337292.337523. 42, 49

[21] Y. K. Choi, J. Cong, and D. Wu. Fpga implementation of em algorithm for 3d ct reconstruction. In *Proceedings of the 2014 IEEE 22Nd International Symposium on Field-Programmable Custom Computing Machines*, FCCM '14, pages 157–160, Washington, DC, USA, 2014. IEEE Computer Society. DOI: 10.1109/FCCM.2014.48. 46

[22] Y. K. Choi, J. Cong, and D. Wu. Fpga implementation of em algorithm for 3d ct reconstruction. In *Field-Programmable Custom Computing Machines (FCCM), 2014 IEEE 22nd Annual International Symposium on*, pages 157–160. IEEE, 2014. DOI: 10.1109/FCCM.2014.48. 87

[23] E. S. Chung, J. D. Davis, and J. Lee. Linqits: Big data on little clients. In *Proceedings of the 40th Annual International Symposium on Computer Architecture*, pages 261–272. ACM, 2013. DOI: 10.1145/2485922.2485945. 30

[24] N. T. Clark, H. Zhong, and S. A. Mahlke. Automated custom instruction generation for domain-specific processor acceleration. *Computers, IEEE Transactions on*, 54(10):1258–1270, 2005. DOI: 10.1109/TC.2005.156. 20, 23

[25] J. Cong, Y. Fan, G. Han, and Z. Zhang. Application-specific instruction generation for configurable processor architectures. In *Proceedings of the 2004 ACM/SIGDA 12th international symposium on Field programmable gate arrays*, pages 183–189. ACM, 2004. DOI: 10.1145/968280.968307. 20, 23

[26] J. Cong, M. A. Ghodrat, M. Gill, B. Grigorian, K. Gururaj, and G. Reinman. Accelerator-rich architectures: Opportunities and progresses. In *Proceedings of the The 51st Annual Design Automation Conference on Design Automation Conference*, pages 1–6. ACM, 2014. DOI: 10.1145/2593069.2596667. 3, 13, 14, 15, 30, 33, 36, 37, 41

[27] J. Cong, M. A. Ghodrat, M. Gill, B. Grigorian, H. Huang, and G. Reinman. Composable accelerator-rich microprocessor enhanced for adaptivity and longevity. In *Low Power Electronics and Design (ISLPED), 2013 IEEE International Symposium on*, pages 305–310. IEEE, 2013. DOI: 10.1109/ISLPED.2013.6629314. 3

[28] J. Cong, M. A. Ghodrat, M. Gill, B. Grigorian, and G. Reinman. Architecture support for accelerator-rich cmps. In *Proceedings of the 49th Annual Design Automation Conference*, pages 843–849. ACM, 2012. DOI: 10.1145/2228360.2228512. 3, 26, 28, 29, 55

[29] J. Cong, M. A. Ghodrat, M. Gill, C. Liu, and G. Reinman. Bin: A buffer-in-nuca scheme for accelerator-rich cmps. In *Proceedings of the 2012 ACM/IEEE International Symposium on Low Power Electronics and Design*, ISLPED '12, pages 225–230, New York, NY, USA, 2012. ACM. DOI: 10.1145/2333660.2333715. 42, 46, 49, 55, 56, 57, 58

[30] J. Cong, M. Gill, Y. Hao, G. Reinman, and B. Yuan. On-chip interconnection network for accelerator-rich architectures. In *Proceedings of the 52th Annual Design Automation Conference*, DAC '15, New York, NY, USA, 2015. ACM. DOI: 10.1145/2744769.2744879. 70

[31] J. Cong, H. Guoling, A. Jagannathan, G. Reinman, and K. Rutkowski. Accelerating sequential applications on cmps using core spilling. *Parallel and Distributed Systems, IEEE Transactions on*, 18(8):1094–1107, 2007. DOI: 10.1109/TPDS.2007.1085. 17

[32] J. Cong, K. Gururaj, H. Huang, C. Liu, G. Reinman, and Y. Zou. An energy-efficient adaptive hybrid cache. In *Proceedings of the 17th IEEE/ACM International Symposium on Low-power Electronics and Design*, ISLPED '11, pages 67–72, Piscataway, NJ, USA, 2011. IEEE Press. DOI: 10.1109/ISLPED.2011.5993609. 42, 49, 51, 52, 53, 54, 62, 65

[33] J. Cong, P. Li, B. Xiao, and P. Zhang. An Optimal Microarchitecture for Stencil Computation Acceleration Based on Non-Uniform Partitioning of Data Reuse Buffers. In *Proceedings of the The 51st Annual Design Automation Conference on Design Automation Conference - DAC '14*, pages 1–6, 2014. DOI: 10.1145/2593069.2593090. 46, 49, 75, 79, 80

[34] J. Cong, P. Li, B. Xiao, and P. Zhang. An Optimal Microarchitecture for Stencil Computation Acceleration Based on Non-Uniform Partitioning of Data Reuse Buffers. Technical report, Computer Science Department, UCLA, TR140009, 2014. DOI: 10.1145/2593069.2593090. 78

[35] J. Cong, B. Liu, S. Neuendorffer, J. Noguera, K. Vissers, and Z. Zhang. High-level synthesis for fpgas: From prototyping to deployment. *Computer-Aided Design of Integrated Circuits and Systems, IEEE Transactions on*, 30(4):473–491, 2011. DOI: 10.1109/T-CAD.2011.2110592. 30

[36] J. Cong, B. Liu, S. Neuendorffer, J. Noguera, K. Vissers, and Z. Zhang. High-Level Synthesis for FPGAs: From Prototyping to Deployment. *IEEE Transactions on Computer-Aided Design of Integrated Circuits and Systems*, 30(4):473–491, Apr. 2011. DOI: 10.1109/TCAD.2011.2110592. 73

[37] J. Cong, C. Liu, and G. Reinman. ACES : Application-Specific Cycle Elimination and Splitting for Deadlock-Free Routing on Irregular Network-on-Chip. In *Proceedings of the 47th Design Automation Conference on - DAC '10*, page 443, 2010. DOI: 10.1145/1837274.1837385. 75, 76, 77, 78

[38] J. Cong, V. Sarkar, G. Reinman, and A. Bui. Customizable Domain-Specific Computing. *IEEE Design and Test of Computers*, 28(2):6–15, Mar. 2011. DOI: 10.1109/MDT.2010.141. 2, 79

[39] J. Cong, V. Sarkar, G. Reinman, and A. Bui. Customizable domain-specific computing. *Design Test of Computers, IEEE*, 28(2):6–15, March 2011. DOI: 10.1109/MDT.2010.141. 5

[40] J. Cong and B. Xiao. Optimization of Interconnects Between Accelerators and Shared Memories in Dark Silicon. In *International Conference on Computer-Aided Design (IC-CAD)*, 2013. DOI: 10.1109/ICCAD.2013.6691182. 73, 74, 75

[41] J. Cong and B. Xiao. FPGA-RPI: A Novel FPGA Architecture With RRAM-Based Programmable Interconnects. *IEEE Transactions on Very Large Scale Integration (VLSI) Systems*, 22(4):864–877, Apr. 2014. DOI: 10.1109/TVLSI.2013.2259512. 82, 83

[42] J. Cong, P. Zhang, and Y. Zou. Combined loop transformation and hierarchy allocation for data reuse optimization. In *Proceedings of the International Conference on Computer-Aided Design*, ICCAD '11, pages 185–192, Piscataway, NJ, USA, 2011. IEEE Press. DOI: 10.1109/ICCAD.2011.6105324. 55

[43] H. Cook, K. Asanović, and D. A. Patterson. Virtual local stores: Enabling software-managed memory hierarchies in mainstream computing environments. Technical Report UCB/EECS-2009-131, EECS Department, University of California, Berkeley, Sep 2009. 42

[44] L. P. Cordella, P. Foggia, C. Sansone, and M. Vento. A (sub) graph isomorphism algorithm for matching large graphs. *Pattern Analysis and Machine Intelligence, IEEE Transactions on*, 26(10):1367–1372, 2004. DOI: 10.1109/TPAMI.2004.75. 23

[45] R. Dennard, F. Gaensslen, V. Rideout, E. Bassous, and A. LeBlanc. Design of ion-implanted MOSFET's with very small physical dimensions. *IEEE Journal of Solid-State Circuits*, 9(5):256–268, Oct. 1974. DOI: 10.1109/JSSC.1974.1050511. 1

[46] P. Dlugosch, D. Brown, P. Glendenning, M. Leventhal, and H. Noyes. An efficient and scalable semiconductor architecture for parallel automata processing. 2014. 87

[47] X. Dong, X. Wu, G. Sun, Y. Xie, H. Li, and Y. Chen. Circuit and microarchitecture evaluation of 3d stacking magnetic ram (mram) as a universal memory replacement. In *Proceedings of the 45th Annual Design Automation Conference*, DAC '08, pages 554–559, New York, NY, USA, 2008. ACM. DOI: 10.1145/1391469.1391610. 58

[48] J. Duato and T. Pinkston. A general theory for deadlock-free adaptive routing using a mixed set of resources. *IEEE Transactions on Parallel and Distributed Systems*, 12(12):1219–1235, 2001. DOI: 10.1109/71.970556. 75

[49] A. E. Eichenberger, K. O'Brien, P. Wu, T. Chen, P. H. Oden, D. A. Prener, J. C. Shepherd, B. So, Z. Sura, A. Wang, et al. Optimizing compiler for the cell processor. In *Parallel Architectures and Compilation Techniques, 2005. PACT 2005. 14th International Conference on*, pages 161–172. IEEE, 2005. DOI: 10.1109/PACT.2005.33. 23

[50] A. E. Eichenberger, P. Wu, and K. O'brien. Vectorization for simd architectures with alignment constraints. In *ACM SIGPLAN Notices*, volume 39, pages 82–93. ACM, 2004. DOI: 10.1145/996893.996853. 23

[51] H. Esmaeilzadeh, E. Blem, R. St. Amant, K. Sankaralingam, and D. Burger. Dark silicon and the end of multicore scaling. In *Proceeding of the 38th annual international symposium on Computer architecture - ISCA '11*, volume 39, pages 365–376, July 2011. DOI: 10.1145/2000064.2000108. 1, 2, 3

[52] R. Espasa, F. Ardanaz, J. Emer, S. Felix, J. Gago, R. Gramunt, I. Hernandez, T. Juan, G. Lowney, M. Mattina, et al. Tarantula: a vector extension to the alpha architecture. In *Computer Architecture, 2002. Proceedings. 29th Annual International Symposium on*, pages 281–292. IEEE, 2002. DOI: 10.1145/545214.545247. 20

[53] C. F. Fajardo, Z. Fang, R. Iyer, G. F. Garcia, S. E. Lee, and L. Zhao. Buffer-integrated-cache: A cost-effective sram architecture for handheld and embedded platforms. In *Proceedings of the 48th Design Automation Conference*, DAC '11, pages 966–971, New York, NY, USA, 2011. ACM. DOI: 10.1145/2024724.2024938. 42, 46, 49, 54, 55

[54] T. Feist. Vivado design suite. *Xilinx, White Paper Version*, 1, 2012. 30

[55] N. Firasta, M. Buxton, P. Jinbo, K. Nasri, and S. Kuo. Intel avx: New frontiers in performance improvements and energy efficiency. *Intel white paper*, 2008. 19, 20

[56] K. Flautner, N. S. Kim, S. Martin, D. Blaauw, and T. Mudge. Drowsy caches: Simple techniques for reducing leakage power. In *Proceedings of the 29th Annual International Symposium on Computer Architecture*, ISCA '02, pages 148–157, Washington, DC, USA, 2002. IEEE Computer Society. DOI: 10.1109/ISCA.2002.1003572. 43

[57] J. Flich and J. Duato. Logic-Based Distributed Routing for NoCs. *IEEE Computer Architecture Letters*, 7(1):13–16, Jan. 2008. DOI: 10.1109/L-CA.2007.16. 76

[58] H. Franke, J. Xenidis, C. Basso, B. M. Bass, S. S. Woodward, J. D. Brown, and C. L. Johnson. Introduction to the wire-speed processor and architecture. *IBM Journal of Research and Development*, 54(1):3–1, 2010. DOI: 10.1147/JRD.2009.2036980. 26, 27, 28

[59] P.-E. Gaillardon, M. Haykel Ben-Jamaa, G. Betti Beneventi, F. Clermidy, and L. Perniola. Emerging memory technologies for reconfigurable routing in FPGA architecture. In *International Conference on Electronics, Circuits and Systems (ICECS)*, pages 62–65, Dec. 2010. DOI: 10.1109/ICECS.2010.5724454. 82

[60] P.-E. Gaillardon, D. Sacchetto, G. B. Beneventi, M. H. Ben Jamaa, L. Perniola, F. Clermidy, I. O'Connor, and G. De Micheli. Design and Architectural Assessment of 3-D Resistive Memory Technologies in FPGAs. *IEEE Transactions on Nanotechnology*, 12(1):40–50, Jan. 2013. DOI: 10.1109/TNANO.2012.2226747. 82

[61] S. Goldstein, H. Schmit, M. Budiu, S. Cadambi, M. Moe, and R. Taylor. PipeRench: a reconfigurable architecture and compiler. *Computer*, 33(4):70–77, Apr. 2000. DOI: 10.1109/2.839324. 3

[62] V. Govindaraju, C.-H. Ho, T. Nowatzki, J. Chhugani, N. Satish, K. Sankaralingam, and C. Kim. DySER: Unifying Functionality and Parallelism Specialization for Energy-Efficient Computing. *IEEE Micro*, 32(5):38–51, Sept. 2012. DOI: 10.1109/MM.2012.51. 3

[63] V. Govindaraju, C.-H. Ho, and K. Sankaralingam. Dynamically specialized datapaths for energy efficient computing. In *High Performance Computer Architecture (HPCA), 2011 IEEE 17th International Symposium on*, pages 503–514. IEEE, 2011. DOI: 10.1109/HPCA.2011.5749755. 19, 21

[64] P. Greenhalgh. Big. little processing with arm cortex-a15 & cortex-a7. *ARM White Paper*, 2011. 17

[65] W. Guan, S. Long, Q. Liu, M. Liu, and W. Wang. Nonpolar Nonvolatile Resistive Switching in Cu Doped ZrO_2. *IEEE Electron Device Letters*, 29(5):434–437, May 2008. DOI: 10.1109/LED.2008.919602. 83

[66] S. Gupta, S. Feng, A. Ansari, S. Mahlke, and D. August. Bundled execution of recurring traces for energy-efficient general purpose processing. In *Proceedings of the 44th Annual IEEE/ACM International Symposium on Microarchitecture*, pages 12–23. ACM, 2011. DOI: 10.1145/2155620.2155623. 19, 21, 22, 23

[67] R. Hameed, W. Qadeer, M. Wachs, O. Azizi, A. Solomatnikov, B. C. Lee, S. Richardson, C. Kozyrakis, and M. Horowitz. Understanding sources of inefficiency in general-purpose chips. *International Symposium on Computer Architecture*, page 37, 2010. DOI: 10.1145/1816038.1815968. 2, 73

[68] T. Hayes, O. Palomar, O. Unsal, A. Cristal, and M. Valero. Vector extensions for decision support dbms acceleration. In *Microarchitecture (MICRO), 2012 45th Annual IEEE/ACM International Symposium on*, pages 166–176. IEEE, 2012. DOI: 10.1109/MICRO.2012.24. 20

[69] B. He, K. Yang, R. Fang, M. Lu, N. Govindaraju, Q. Luo, and P. Sander. Relational joins on graphics processors. In *Proceedings of the 2008 ACM SIGMOD international conference on Management of data*, pages 511–524. ACM, 2008. DOI: 10.1145/1376616.1376670. 87

[70] T. Henretty, J. Holewinski, N. Sedaghati, L.-N. Pouchet, A. Rountev, and P. Sadayappan. Stencil Domain Specific Language (SDSL) User Guide 0.2.1 draft. Technical report, OSU TR OSU-CISRC-4/13-TR09, 2013. 78

[71] H. P. Hofstee. Power efficient processor architecture and the cell processor. In *High-Performance Computer Architecture, 2005. HPCA-11. 11th International Symposium on*, pages 258–262. IEEE, 2005. DOI: 10.1109/HPCA.2005.26. 17

[72] IBM. Power8 coherent accelerator processor interface (CAPI). 86

[73] Intel. Intel QuickAssist acceleration technology for embedded systems. 86

[74] E. Ipek, M. Kirman, N. Kirman, and J. F. Martinez. Core fusion: accommodating software diversity in chip multiprocessors. In *ACM SIGARCH Computer Architecture News*, volume 35, pages 186–197. ACM, 2007. DOI: 10.1145/1273440.1250686. 17, 18

[75] I. Issenin, E. Brockmeyer, M. Miranda, and N. Dutt. Drdu: A data reuse analysis technique for efficient scratch-pad memory management. *ACM Trans. Des. Autom. Electron. Syst.*, 12(2), Apr. 2007. DOI: 10.1145/1230800.1230807. 41

[76] A. Jadidi, M. Arjomand, and H. Sarbazi-Azad. High-endurance and performance-efficient design of hybrid cache architectures through adaptive line replacement. In *Low Power Electronics and Design (ISLPED) 2011 International Symposium on*, pages 79–84, Aug 2011. DOI: 10.1109/ISLPED.2011.5993611. 42, 58, 59, 60, 61, 63, 64, 65

[77] S. Kaxiras, Z. Hu, and M. Martonosi. Cache decay: Exploiting generational behavior to reduce cache leakage power. In *Proceedings of the 28th Annual International Symposium on Computer Architecture*, ISCA '01, pages 240–251, New York, NY, USA, 2001. ACM. DOI: 10.1109/ISCA.2001.937453. 41, 43, 45, 46, 47, 60, 61

[78] S. Kaxiras and M. Martonosi. *Computer Architecture Techniques for Power-Efficiency.* Morgan and Claypool Publishers, 1st edition, 2008. DOI: 10.2200/S00119ED1V01Y200805CAC004. 43

[79] Y. Kim, G.-S. Byun, A. Tang, C.-P. Jou, H.-H. Hsieh, G. Reinman, J. Cong, and M. Chang. An 8gb/s/pin 4pj/b/pin single-t-line dual (base+rf) band simultaneous bidirectional mobile memory i/o interface with inter-channel interference suppression. In *Solid-State Circuits Conference Digest of Technical Papers (ISSCC), 2012 IEEE International*, pages 50–52. IEEE, 2012. DOI: 10.1109/ISSCC.2012.6176874. 87

[80] T. Kluter, P. Brisk, P. Ienne, and E. Charbon. Way stealing: Cache-assisted automatic instruction set extensions. In *Proceedings of the 46th Annual Design Automation Conference*, DAC '09, pages 31–36, New York, NY, USA, 2009. ACM. DOI: 10.1145/1629911.1629923. 42

[81] M. Koester, M. Porrmann, and H. Kalte. Task placement for heterogeneous reconfigurable architectures. In *Field-Programmable Technology, 2005. Proceedings. 2005 IEEE International Conference on*, pages 43–50. IEEE, 2005. 32

[82] M. Kong, R. Veras, K. Stock, F. Franchetti, L.-N. Pouchet, and P. Sadayappan. When polyhedral transformations meet simd code generation. *ACM SIGPLAN Notices*, 48(6):127–138, 2013. DOI: 10.1145/2499370.2462187. 23

[83] A. Kumar, L.-S. Peh, P. Kundu, and N. K. Jha. Express virtual channels: Towards the ideal interconnection fabric. In *Proceedings of the 34th Annual International Symposium on Computer Architecture*, ISCA '07, pages 150–161, New York, NY, USA, 2007. ACM. DOI: 10.1145/1273440.1250681. 70

[84] R. Kumar, K. I. Farkas, N. P. Jouppi, P. Ranganathan, and D. M. Tullsen. Single-isa heterogeneous multi-core architectures: The potential for processor power reduction. In *Microarchitecture, 2003. MICRO-36. Proceedings. 36th Annual IEEE/ACM International Symposium on*, pages 81–92. IEEE, 2003. DOI: 10.1145/956417.956569. 17

[85] I. Kuon and J. Rose. Measuring the Gap Between FPGAs and ASICs. *IEEE Transactions on Computer-Aided Design of Integrated Circuits and Systems*, 26(2):203–215, Feb. 2007. DOI: 10.1109/TCAD.2006.884574. 86

[86] B. C. Lee, E. Ipek, O. Mutlu, and D. Burger. Architecting phase change memory as a scalable dram alternative. In *Proceedings of the 36th Annual International Symposium on Computer Architecture*, ISCA '09, pages 2–13, New York, NY, USA, 2009. ACM. DOI: 10.1145/1555815.1555758. 58

[87] P. Li, Y. Wang, P. Zhang, G. Luo, T. Wang, and J. Cong. Memory partitioning and scheduling co-optimization in behavioral synthesis. In *International Conference on Computer-Aided Design*, pages 488–495, 2012. DOI: 10.1145/2429384.2429484. 78

[88] M. Lipson. Guiding, modulating, and emitting light on Silicon-challenges and opportunities. *Journal of Lightwave Technology*, 23(12):4222–4238, Dec. 2005. DOI: 10.1109/JLT.2005.858225. 81

[89] M. J. Lyons, M. Hempstead, G.-Y. Wei, and D. Brooks. The accelerator store: A shared memory framework for accelerator-based systems. *ACM Trans. Archit. Code Optim.*, 8(4):48:1–48:22, Jan. 2012. DOI: 10.1145/2086696.2086727. 3, 41, 42, 46, 47, 48, 73

[90] J. D. C. Maia, G. A. Urquiza Carvalho, C. P. Mangueira Jr, S. R. Santana, L. A. F. Cabral, and G. B. Rocha. Gpu linear algebra libraries and gpgpu programming for accelerating mopac semiempirical quantum chemistry calculations. *Journal of Chemical Theory and Computation*, 8(9):3072–3081, 2012. DOI: 10.1021/ct3004645. 87

[91] A. Marshall, T. Stansfield, I. Kostarnov, J. Vuillemin, and B. Hutchings. A reconfigurable arithmetic array for multimedia applications. In *International Symposium on FPGAs*, pages 135–143, 1999. DOI: 10.1145/296399.296444. 3

[92] B. Mei, S. Vernalde, D. Verkest, H. De Man, and R. Lauwereins. Exploiting loop-level parallelism on coarse-grained reconfigurable architectures using modulo scheduling. In *Computers and Digital Techniques, IEE Proceedings-*, volume 150, pages 255–61. IET, 2003. DOI: 10.1049/ip-cdt:20030833. 30

[93] A. Meyerson and B. Tagiku. *Approximation, Randomization, and Combinatorial Optimization. Algorithms and Techniques*, volume 5687 of *Lecture Notes in Computer Science*. Springer Berlin Heidelberg, Berlin, Heidelberg, 2009. 69

[94] E. Mirsky and A. Dehon. MATRIX: a reconfigurable computing architecture with configurable instruction distribution and deployable resources. In *IEEE Symposium on FPGAs for Custom Computing Machines*, pages 157–166, 1996. DOI: 10.1109/FPGA.1996.564808. 3

[95] R. K. Montoye, E. Hokenek, and S. L. Runyon. Design of the ibm risc system/6000 floating-point execution unit. *IBM Journal of research and development*, 34(1):59–70, 1990. DOI: 10.1147/rd.341.0059. 19, 20

[96] C. A. Moritz, M. I. Frank, and S. Amarasinghe. Flexcache: A framework for flexible compiler generated data caching, 2001. DOI: 10.1007/3-540-44570-6_9. 42

[97] A. A. Nacci, V. Rana, F. Bruschi, D. Sciuto, I. Beretta, and D. Atienza. A high-level synthesis flow for the implementation of iterative stencil loop algorithms on FPGA devices. In *Design Automation Conference*, page 1, 2013. DOI: 10.1145/2463209.2488797. 79

[98] U. Nawathe, M. Hassan, L. Warriner, K. Yen, B. Upputuri, D. Greenhill, A. Kumar, and H. Park. An 8-core, 64-thread, 64-bit, power efficient sparc soc (niagara 2). *ISSCC, http://www. opensparc. net/pubs/preszo/07/n2isscc. pdf*, 2007. DOI: 10.1145/1231996.1232000. 26

[99] U. Ogras and R. Marculescu. Energy- and Performance-Driven NoC Communication Architecture Synthesis Using a Decomposition Approach. In *Design, Automation and Test in Europe*, number 9097, pages 352–357, 2005. DOI: 10.1109/DATE.2005.137. 69, 70

[100] U. Ogras and R. Marculescu. "It's a small world after all": NoC performance optimization via long-range link insertion. *IEEE Transactions on Very Large Scale Integration (VLSI) Systems*, 14(7):693–706, July 2006. DOI: 10.1109/TVLSI.2006.878263. 69, 76

[101] S. Onkaraiah, P.-e. Gaillardon, M. Reyboz, F. Clermidy, J.-m. Portal, M. Bocquet, and C. Muller. Using OxRRAM memories for improving communications of reconfigurable FPGA architectures. In *International Symposium on Nanoscale Architectures (NANOARCH)*, pages 65–69, June 2011. DOI: 10.1109/NANOARCH.2011.5941485. 82

[102] J. Ouyang, S. Lin, Z. Hou, P. Wang, Y. Wang, and G. Sun. Active ssd design for energy-efficiency improvement of web-scale data analysis. In *Proceedings of the International Symposium on Low Power Electronics and Design*, pages 286–291. IEEE Press, 2013. DOI: 10.1109/ISLPED.2013.6629310. 87

[103] M. Palesi, R. Holsmark, S. Kumar, and V. Catania. Application Specific Routing Algorithms for Networks on Chip. *IEEE Transactions on Parallel and Distributed Systems*, 20(3):316–330, Mar. 2009. DOI: 10.1109/TPDS.2008.106. 75, 76

[104] H. Park, Y. Park, and S. Mahlke. Polymorphic pipeline array: a flexible multicore accelerator with virtualized execution for mobile multimedia applications. In *Proceedings of the 42nd Annual IEEE/ACM International Symposium on Microarchitecture*, pages 370–380. ACM, 2009. DOI: 10.1145/1669112.1669160. 30, 32

[105] D. Pham, T. Aipperspach, D. Boerstler, M. Bolliger, R. Chaudhry, D. Cox, P. Harvey, P. Harvey, H. Hofstee, C. Johns, J. Kahle, A. Kameyama, J. Keaty, Y. Masubuchi, M. Pham, J. Pille, S. Posluszny, M. Riley, D. Stasiak, M. Suzuoki, O. Takahashi, J. Warnock, S. Weitzel, D. Wendel, and K. Yazawa. Overview of the architecture, circuit design, and physical implementation of a first-generation cell processor. *Solid-State Circuits, IEEE Journal of*, 41(1):179–196, Jan 2006. DOI: 10.1109/JSSC.2005.859896. 39, 42

[106] A. Pinto, L. Carloni, and A. Sangiovanni-Vincentelli. Efficient synthesis of networks on chip. In *Proceedings 21st International Conference on Computer Design*, pages 146–150, 2003. DOI: 10.1109/ICCD.2003.1240887. 69, 70

[107] M. Powell, S.-H. Yang, B. Falsafi, K. Roy, and T. N. Vijaykumar. Gated-vdd: A circuit technique to reduce leakage in deep-submicron cache memories. In *Proceedings of the 2000 International Symposium on Low Power Electronics and Design*, ISLPED '00, pages 90–95, New York, NY, USA, 2000. ACM. DOI: 10.1145/344166.344526. 41, 43, 45, 60, 61

[108] M. Pricopi and T. Mitra. Bahurupi: A polymorphic heterogeneous multi-core architecture. *ACM Transactions on Architecture and Code Optimization (TACO)*, 8(4):22, 2012. DOI: 10.1145/2086696.2086701. 17

[109] A. Putnam, A. M. Caulfield, E. S. Chung, D. Chiou, K. Constantinides, J. Demme, H. Esmaeilzadeh, J. Fowers, G. P. Gopal, J. Gray, et al. A reconfigurable fabric for accelerating large-scale datacenter services. In *Computer Architecture (ISCA), 2014 ACM/IEEE 41st International Symposium on*, pages 13–24. IEEE, 2014. DOI: 10.1109/ISCA.2014.6853195. 87

[110] M. K. Qureshi, D. Thompson, and Y. N. Patt. The v-way cache: Demand based associativity via global replacement. In *Proceedings of the 32Nd Annual International Symposium on Computer Architecture*, ISCA '05, pages 544–555, Washington, DC, USA, 2005. IEEE Computer Society. DOI: 10.1145/1080695.1070015. 51

[111] S. K. Raman, V. Pentkovski, and J. Keshava. Implementing streaming simd extensions on the pentium iii processor. *IEEE micro*, 20(4):47–57, 2000. DOI: 10.1109/40.865866. 19, 20

[112] A. Ramirez, F. Cabarcas, B. Juurlink, M. Alvarez Mesa, F. Sanchez, A. Azevedo, C. Meenderinck, C. Ciobanu, S. Isaza, and G. Gaydadjiev. The sarc architecture. *IEEE micro*, 30(5):16–29, 2010. DOI: 10.1109/MM.2010.79. 26

[113] P. Ranganathan, S. Adve, and N. P. Jouppi. Reconfigurable caches and their application to media processing. In *Proceedings of the 27th Annual International Symposium on Computer Architecture*, ISCA '00, pages 214–224, New York, NY, USA, 2000. ACM. DOI: 10.1145/342001.339685. 42, 49, 50, 51

[114] R. Riedlinger, R. Bhatia, L. Biro, B. Bowhill, E. Fetzer, P. Gronowski, and T. Grutkowski. A 32nm 3.1 billion transistor 12-wide-issue itanium processor for mission-critical servers. In *Solid-State Circuits Conference Digest of Technical Papers (ISSCC), 2011 IEEE International*, pages 84–86, Feb 2011. DOI: 10.1109/ISSCC.2011.5746230. 43

[115] R. Rodrigues, A. Annamalai, I. Koren, S. Kundu, and O. Khan. Performance per watt benefits of dynamic core morphing in asymmetric multicores. In *Parallel Architectures and Compilation Techniques (PACT), 2011 International Conference on*, pages 121–130. IEEE, 2011. DOI: 10.1109/PACT.2011.18. 17

[116] P. Schaumont and I. Verbauwhede. Domain-specific codesign for embedded security. *Computer*, 36(4):68–74, Apr. 2003. DOI: 10.1109/MC.2003.1193231. 2

[117] S.-S. Sheu, P.-C. Chiang, W.-P. Lin, H.-Y. Lee, P.-S. Chen, T.-Y. Wu, F. T. Chen, K.-L. Su, M.-J. Kao, and K.-H. Cheng. A 5ns Fast Write Multi-Level Non-Volatile 1 K bits RRAM Memory with Advance Write Scheme. In *VLSI Circuits, Symposium on*, pages 82–83, 2009. 83

[118] H. Singh, M.-H. Lee, G. Lu, F. J. Kurdahi, N. Bagherzadeh, and E. M. Chaves Filho. Morphosys: an integrated reconfigurable system for data-parallel and computation-intensive applications. *Computers, IEEE Transactions on*, 49(5):465–481, 2000. DOI: 10.1109/12.859540. 3, 32, 33

[119] A. Solomatnikov, A. Firoozshahian, W. Qadeer, O. Shacham, K. Kelley, Z. Asgar, M. Wachs, R. Hameed, and M. Horowitz. Chip multi-processor generator. In *Proceedings of the 44th annual conference on Design automation - DAC '07*, page 262, 2007. DOI: 10.1145/1278480.1278544. 2

[120] D. Starobinski, M. Karpovsky, and L. Zakrevski. Application of network calculus to general topologies using turn-prohibition. *IEEE/ACM Transactions on Networking*, 11(3):411–421, June 2003. DOI: 10.1109/TNET.2003.813040. 75

[121] G. Sun, X. Dong, Y. Xie, J. Li, and Y. Chen. A novel architecture of the 3d stacked mram l2 cache for cmps. In *High Performance Computer Architecture, 2009. HPCA 2009. IEEE 15th International Symposium on*, pages 239–249, Feb 2009. DOI: 10.1109/HPCA.2009.4798259. 42, 58, 59

[122] S. Tanachutiwat, M. Liu, and W. Wang. FPGA Based on Integration of CMOS and RRAM. *IEEE Transactions on Very Large Scale Integration (VLSI) Systems*, 19(11):2023–2032, Nov. 2011. DOI: 10.1109/TVLSI.2010.2063444. 82

[123] D. Tarjan, M. Boyer, and K. Skadron. Federation: Repurposing scalar cores for out-of-order instruction issue. In *Proceedings of the 45th annual Design Automation Conference*, pages 772–775. ACM, 2008. DOI: 10.1145/1391469.1391666. 17

[124] K. Tsunoda, K. Kinoshita, H. Noshiro, Y. Yamazaki, T. Iizuka, Y. Ito, A. Takahashi, A. Okano, Y. Sato, T. Fukano, M. Aoki, and Y. Sugiyama. Low Power and High Speed Switching of Ti-doped NiO ReRAM under the Unipolar Voltage Source of less than 3V. In *International Electron Devices Meeting (IEDM)*, pages 767–770, Dec. 2007. DOI: 10.1109/IEDM.2007.4419060. 83

[125] O. S. Unsal, C. M. Krishna, and C. Mositz. Cool-fetch: Compiler-enabled power-aware fetch throttling. *Computer Architecture Letters*, 1(1):5–5, 2002. DOI: 10.1109/L-CA.2002.3. 16

[126] D. Vantrease, N. Binkert, R. Schreiber, and M. H. Lipasti. Light speed arbitration and flow control for nanophotonic interconnects. *Proceedings of the 42nd Annual IEEE/ACM International Symposium on Microarchitecture - Micro-42*, page 304, 2009. DOI: 10.1145/1669112.1669152. 69

[127] D. Vantrease, R. Schreiber, M. Monchiero, M. McLaren, N. P. Jouppi, M. Fiorentino, A. Davis, N. Binkert, R. G. Beausoleil, and J. H. Ahn. Corona : System Implications of Emerging Nanophotonic Technology. *2008 International Symposium on Computer Architecture*, pages 153–164, June 2008. DOI: 10.1109/ISCA.2008.35. 81

[128] G. Venkatesh, J. Sampson, N. Goulding, S. Garcia, V. Bryksin, J. Lugo-Martinez, S. Swanson, and M. B. Taylor. Conservation cores: reducing the energy of mature computations. In *ACM SIGARCH Computer Architecture News*, volume 38, pages 205–218. ACM, 2010. DOI: 10.1145/1735970.1736044. 1, 3

[129] C.-H. Wang, Y.-H. Tsai, K.-C. Lin, M.-F. Chang, Y.-C. King, C.-J. Lin, S.-S. Sheu, Y.-S. Chen, H.-Y. Lee, F. T. Chen, and M.-J. Tsai. Three-Dimensional $4F^2$ ReRAM Cell with CMOS Logic Compatible Process. In *IEDM Technical Digest*, pages 664–667, 2010. DOI: 10.1109/IEDM.2010.5703446. 83

[130] Y. Wang, P. Li, P. Zhang, C. Zhang, and J. Cong. Memory partitioning for multidimensional arrays in high-level synthesis. In *Design Automation Conference*, page 1, 2013. DOI: 10.1145/2463209.2488748. 78

[131] Z. Wang, D. Jimenez, C. Xu, G. Sun, and Y. Xie. Adaptive placement and migration policy for an stt-ram-based hybrid cache. In *High Performance Computer Architecture (HPCA), 2014 IEEE 20th International Symposium on*, pages 13–24, Feb 2014. DOI: 10.1109/H-PCA.2014.6835933. 42, 58, 59, 60, 61, 63, 66, 67

[132] X. Wu, J. Li, L. Zhang, E. Speight, R. Rajamony, and Y. Xie. Hybrid cache architecture with disparate memory technologies. In *Proceedings of the 36th Annual International Symposium on Computer Architecture*, ISCA '09, pages 34–45, New York, NY, USA, 2009. ACM. DOI: 10.1145/1555815.1555761. 42, 43, 58, 59, 60, 61, 63

[133] Xilinx. Zynq-7000 all programmable soc. 86

[134] S.-H. Yang, B. Falsafi, M. D. Powell, and T. N. Vijaykumar. Exploiting choice in resizable cache design to optimize deep-submicron processor energy-delay. In *Proceedings of the 8th International Symposium on High-Performance Computer Architecture*, HPCA '02, pages 151–, Washington, DC, USA, 2002. IEEE Computer Society. DOI: 10.1109/H-PCA.2002.995706. 43

[135] Y. Ye, S. Borkar, and V. De. A new technique for standby leakage reduction in high-performance circuits. In *VLSI Circuits, 1998. Digest of Technical Papers. 1998 Symposium on*, pages 40–41, June 1998. DOI: 10.1109/VLSIC.1998.687996. 44

[136] P. Yu and T. Mitra. Scalable custom instructions identification for instruction-set extensible processors. In *Proceedings of the 2004 international conference on Compilers, architecture, and synthesis for embedded systems*, pages 69–78. ACM, 2004. DOI: 10.1145/1023833.1023844. 19, 20, 23

[137] M. Zhang and K. Asanović. Fine-grain cam-tag cache resizing using miss tags. In *Proceedings of the 2002 International Symposium on Low Power Electronics and Design*, ISLPED '02, pages 130–135, New York, NY, USA, 2002. ACM. DOI: 10.1109/LPE.2002.146725. 52, 62, 65

Authors' Biographies

YU-TING CHEN

Yu-Ting Chen is a Ph.D. candidate in the Computer Science Department at the University of California, Los Angeles. He received a B.S. degree in computer science, a B.A. degree in economics, and an M.S. degree in computer science from National Tsing Hua University, HsinTsu, Taiwan, R.O.C., in 2005 and 2007, respectively. He worked at TSMC as a summer intern in 2005 and at Intel Labs as a summer intern in 2013. His research interests include computer architecture, cluster computing, and bioinformatics in DNA sequencing technologies.

JASON CONG

Jason Cong received his B.S. degree in computer science from Peking University in 1985, and his M.S. and Ph.D. degrees in computer science from the University of Illinois at Urbana–Champaign in 1987 and 1990, respectively. Currently, he is a Chancellors Professor at the Computer Science Department, with a joint appointment from the Electrical Engineering Department, at the University of California, Los Angeles. He is the director of the Center for Domain-Specific Computing (CDSC), co-director of the UCLA/Peking University Joint Research Institute in Science and Engineering, and director of the VLSI Architecture, Synthesis, and Technology (VAST) Laboratory. He also served as the chair the UCLA Computer Science Department from 2005–2008. Dr. Cong's research interests include synthesis of VLSI circuits and systems, programmable systems, novel computer architectures, nano-systems, and highly scalable algorithms. He has over 400 publications in these areas, including 10 best paper awards, two 10-Year Most Influential Paper Awards (from ICCAD'14 and ASPDAC'15), and the 2011 ACM/IEEE A. Richard Newton Technical Impact Award in Electric Design Automation. He was elected to an IEEE Fellow in 2000 and ACM Fellow in 2008. He is the recipient of the 2010 IEEE Circuits and System (CAS) Society Technical Achievement Award "for seminal contributions to electronic design automation, especially in FPGA synthesis, VLSI interconnect optimization, and physical design automation."

MICHAEL GILL

Michael Gill received a B.S. degree in computer science from California Polytechnic University, Pomona, and an M.S. and a Ph.D. in computer science from the University of California, Los

Angeles. His research is primarily focused on high-performance architectures, and the interaction between these architectures and compilers, run time systems, and operating systems.

GLENN REINMAN

Glenn Reinman received his B.S. in computer science and engineering from the Massachusetts Institute of Technology in 1996. He earned his M.S. and Ph.D in computer science from the University of California, San Diego, in 1999 and 2001, respectively. He is currently a professor in the Computer Science Department at the University of California, Los Angeles.

BINGJUN XIAO

Bingjun Xiao received a B.S. degree in microelectronics from Peking University, Beijing, China, in 2010. He received an M.S. degree and a Ph.D. degree in electrical engineering from UCLA in 2012 and 2015, respectively. His research interests include machine learning, cluster computing, and data flow optimization.